*The control of gene expression
in animal development*

The control of gene expression in animal development

J. B. GURDON

MEDICAL RESEARCH COUNCIL,
LABORATORY OF MOLECULAR BIOLOGY,
CAMBRIDGE

CLARENDON PRESS · OXFORD

1974

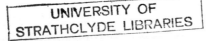

Oxford University Press, Ely House, London W. 1

GLASGOW NEW YORK TORONTO MELBOURNE WELLINGTON
CAPE TOWN IBADAN NAIROBI DAR ES SALAAM LUSAKA ADDIS ABABA
DELHI BOMBAY CALCUTTA MADRAS KARACHI LAHORE DACCA
KUALA LUMPUR SINGAPORE HONG KONG TOKYO

CASEBOUND ISBN O 19 857376 6
PAPERBACK ISBN O 19 857387 1

© OXFORD UNIVERSITY PRESS 1974

PRINTED IN GREAT BRITAIN
BY WESTERN PRINTING SERVICES LIMITED
BRISTOL ENGLAND

Preface

THE main part of this book (Chapters 1–3) originated as the Dunham Lectures given to the Harvard Medical School in November 1971. As presented now, the content of these lectures has been brought up to date, and expanded in some parts. The aim has been to make the subject matter, which is not complicated, intelligible to as wide a range of biologists as possible. With this in mind a glossary and diagram of normal development are included at the end of the book.

Each of the main chapters starts with a general review of background information. The last half of each chapter includes a more detailed account of some recent or current experiments. An attempt has been made to include the more technical or specialized information in Tables or Figures which will not necessarily be understood by those without previous knowledge of cell biology.

Much emphasis has been placed in this book on the use of manipulative experiments on amphibian eggs, for reasons explained in the Introduction. In view of this it was felt useful to include some technical information in an Appendix, which also contains comments on some applications of the techniques described in the text.

J. B. GURDON
Cambridge
July 1973.

Acknowledgements

I AM greatly indebted to my present and past colleagues with whom it has been a pleasure to work, and from whom I have learnt so much. Some of these have contributed very significantly to the original work described in this book, and the following have kindly criticized parts of the manuscript or permitted me to quote recent results: R. M. Benbow, W. M. Bonner, C. C. Ford, J. S. Knowland, C. D. Lane, R. A. Laskey, J. B. Lingrel, R. Reeves, and H. R. Woodland. In addition, I am grateful to Andrew Murray of the Perse School, Cambridge for comments on the manuscript.

I am very glad to acknowledge the tenure of a Research Fellowship at Christ Church, Oxford, which greatly improved circumstances for my experimental work from 1962–72.

Above all I wish to thank my parents without whose help and encouragement it would not have been possible for me to study Biology in the first instance.

Contents

Abbreviations

dATP, dCTP, dGTP, dTTP: deoxytriphosphates of adenosine, cytidine, guanosine, and thymidine.

mRNA: messenger RNA or 'message'.

rRNA: ribosomal RNA.

28s rRNA ⎫
18s rRNA ⎬ three major kinds of ribosomal RNA
5s rRNA ⎭ present as one molecule per ribosome; see Fig. 2.

4s RNA: low molecular weight RNA, often containing tRNA.

tRNA: transfer RNA, which conveys amino acids to ribosomes for assembly into proteins: there are one or more kinds of tRNA for each of the 20 amino acids.

met-tRNA^{met}: transfer RNA of the kind that conveys methionine to positions in the middle of protein molecules.

met-tRNA^{fmet}: transfer RNA of a special kind that conveys methionine to the first position of a new protein.

EMC: encephalomyocarditis virus; a small virus which contains a single-stranded RNA and which normally causes lesions of the nervous system and muscles in mice.

Hb: haemoglobin.

mRNP: messenger RNA complexed with protein; this complex is obtained by versene treatment of polysomes.

ng: nanograms (10^{-9} grams).

pg: picograms (10^{-12} grams).

poly-A: a sequence of up to 200 adjacent polyadenylic acid residues which commonly exist near one end of messenger RNA molecules.

F_1: the first generation of animals to be derived from a cross.

F_2: the second generation of animals to be derived from a cross; the F_2 generation results from crossing two F_1 individuals.

Introduction

How is gene expression controlled, and what determines the distribution of materials in and among cells? These are the two most important questions in animal development. An answer to the first question would explain how the various cell-types of an individual synthesize different proteins and hence products of different enzyme-controlled reactions. At present rather little can be said on the second question, and this book is concerned only with the first.

Chapters 1–3 deal with the three major levels at which gene expression can be controlled. Control at the level of DNA synthesis (Chapter 1) could result in the unequal replication of genes so that there would be differences in the gene content of cells in the same individual. Control at the level of RNA synthesis (Chapter 3) would result in the synthesis of different kinds of messenger RNA, and hence proteins, by cells which have the same genes. Lastly, different cells could have the same genes, all of which are transcribed into messenger RNA at the same rate, but control at the level of protein synthesis (or message translation) could result in different proteins being synthesized in each cell-type. Control at this level is discussed in Chapter 2.

The discussion of gene expression in this book is biased in two directions—towards the use of microinjection as an experimental method, and the use of Amphibia as a source of experimental material. These are the two aspects of developmental research in which the author has been most closely involved, and the lectures which formed the basis of this book encouraged emphasis in areas of personal competence. However,

quite apart from this, work in these areas is of central importance in our present understanding of animal development.

Micromanipulation, biochemistry, and genetics are the three
experimental methods of greatest current value in the analysis
of animal development. Microinjection is an aspect of manipulation which has come to be used commonly in embryological
research only during the last 10–20 years, and there is at
present no full discussion of the advantages and limitations of
this technique. In the author's opinion, manipulation, and
especially microinjection, can play an increasingly useful role
in the analysis of control processes in *living* cells. In accordance
with an emphasis on manipulation, no attempt is made to give
a full account of current biochemical work on development,
though this is referred to where appropriate.

Amphibia have for many years been favoured for embryological research on the grounds of the large size and ready
availability of their eggs and embryos. For microinjection,
amphibian eggs are preferable to those of other species, since
their large size and resistance to penetration enables relatively
large amounts of material to be inserted into each cell. *Xenopus
laevis*, a South African frog which is permanently aquatic, is
particularly valuable for embryological work since large
numbers of eggs can be obtained throughout the year by the
injection of crude preparations of mammalian gonadotropic
hormones. Furthermore this species can be reared to maturity
in the laboratory within a year, so that use can be made of
genetic mutants. Amphibian eggs and embryos are therefore
better than those of any other group for manipulative experiments, at least as good as any others for biochemical work, and
usable from a genetic point of view. This combination of
advantages is responsible for the fact that the development of
amphibia, and in particular of *Xenopus laevis*, is better understood, at present, than that of any other animal species.

1 *Nuclear transplantation and somatic cell genes*

1.1. Is the informational content of chromosomal DNA the same in all somatic cells ?

ANY departure from the normal pattern of gene replication and distribution during development could lead to the cells of an individual having different combinations of genes. Permanent changes in the genome of somatic cells could account for the different populations of gene products observed in specialized somatic cells. The principal changes which the genome might undergo fall into three categories (Fig. 1).

Before discussing reasons for or against such changes, it is useful to consider methods by which the gene content of chromosomal DNA can be investigated in somatic cells. Conventional genetic analysis is not informative; the fact that the progeny of a mating between two animals develop normally proves that the gene content of germ-line cells (those that give rise to sperm and eggs) is not permanently changed during development, but the gene content of somatic cells (all cells other than the germ-line) is not tested in this way. Cytological or biochemical methods can be used to compare directly the DNA of different kinds of specialized cells. Alternatively, the cells or genome of a specialized tissue can be propagated under conditions which cause daughter cells to specialize in different ways, and 'developmental' evidence can then be obtained for the conservation of genes which were unexpressed in the original genome. Direct biochemical or cytological measurements have the advantage of being able to reveal, in an adult tissue, any temporary variations in DNA composition that might be reversed by mitosis as well as permanent changes, but such measurements are not very sensitive. In contrast, the developmental approach has immense sensitivity capable of revealing single gene changes almost anywhere in the genome. In this chapter, biochemical and cytological experiments are covered in an introductory way; the developmental approach is discussed in detail in the rest of the chapter.

FIG. 1 Three categories of stable DNA changes affecting gene expression. The thin line represents part of a chromosome containing three genes, A, B, C. Type I involves a change in the relative position of a gene; type II is a quantitative change affecting the numbers of copies of a gene; type III includes qualitative changes affecting the composition of a gene. Although the permanent inactivation of a gene does not necessarily involve a change in its composition, this type of change is included in the scheme because it would be revealed by the same design of experiment as are gene mutations and deletions in somatic cells.

1.1.1. *Changes in the relative position of genes*

It is firmly established that the relative position of genes in a chromosome set can substantially influence their expression. This effect is clearly seen when chromosomal translocations or inversions place a gene next to a particular (heterochromatic) region of a chromosome, as a result of which its activity (judged by lack of transcription and delayed replication) is decreased (e.g. Brown 1966). Similarly McClintock (1956) has described, in maize, some 'transposible elements'; though it is not clear whether these elements are genes in the sense of containing DNA and serving as a template for transcription, they can move from one chromosome to another and can affect genes in whose proximity they come to lie by making them hyper-mutable.

The question of whether changes of gene sequence are likely to be important in development is best answered at present by noting that whenever the phenomenon is known to take place, it does so irregularly. This is not what would be expected of a normal developmental mechanism except in the special case of antibody formation where substantial variability in the arrangement of certain sequences could be useful in bringing about the variable composition of immunoglobulins.

In summary, we can say that changes in gene sequence can affect gene expression, that phenomena which resemble changes in gene location have been observed, but that this is unlikely to be a generally important developmental mechanism because of its irregular occurrence.

1.1.2. *Changes in the numbers of genes*

An equal increase of all genes in a genome, as occurs in polyploidy, is most unlikely to alter the balance of gene expression, and is not discussed further. Gene amplification refers to an increase in the number of one or more genes in the genome without a proportionate increase in others. It is known that some genes are present in many copies in a chromosome set (and are in this sense amplified), while others are probably present in only one copy per haploid set of chromosomes. From a developmental point of view, we are interested only in the possibility that the same kind of gene is amplified to an unequal extent in the different cell-types of an individual.

Proof of a specific gene amplification of this kind was first provided by Brown and Dawid (1968) and by Gall (1968), for ribosomal DNA (the genes which code for 18s and 28s ribosomal RNA) in *Xenopus laevis*. The proof depended on the biophysical isolation of ribosomal genes (review by Birnstiel *et al.* 1971), so permitting their identification and quantitation, as well as on the recognition of nucleolar DNA at the cytological level. The subject has been reviewed by Gall (1969) and Macgregor (1972). The principal facts are these. In *Xenopus laevis*, each haploid genome contains about 500 adjacent *sets* of ribosomal genes (Fig. 2). During an early stage in meiois

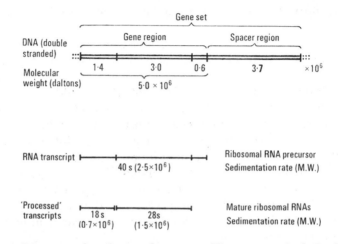

FIG. 2 Diagram of a ribosomal gene set. The gene set includes DNA which codes for 18s and 28s ribosomal RNA, as well as an untranscribed 'spacer' region. In *Xenopus* there are about 500 of these gene sets adjacent to each other on one chromosome, at a site called the nucleolus organizer. This diagram is based on results obtained in *Xenopus* by Birnstiel *et al.* (1971), Dawid *et al.* (1970), Brown and Weber (1968), and Wensinck and Brown (1971).

(pachytene), these gene sets are selectively increased so that each mature oocyte nucleus contains about two million ribosomal gene sets. If we assume that each gene set is used in the amplification process, this would involve a thousand-fold amplification of all gene sets contained in an early oocyte, which, being tetraploid at this stage of meiosis, contains 2000 gene sets (Table 1). The extra ribosomal gene sets are accomodated in some 1000–1500 free nucleoli which can be seen in the nuclei of growing and mature oocytes (Fig. 3c).

Another case in which gene amplification appears to take place is in the polytene chromosomes of salivary glands in certain species of fly larvae. In several members of the family Sciaridae, DNA synthesis is observed at certain chromosomal

TABLE 1

Ribosomal genes of Xenopus laevis in somatic cells and after amplification in oocytes

Cell-type	Nuclear DNA (pg)	Number of chromosome sets per nucleus	Chromosomal DNA (pg per nucleus)	Ribosomal DNA[†] (pg per nucleus)	Number of rDNA[†] gene sets per nucleus	Number of nucleoli
Somatic	6	2	6	0·012 (0·2 per cent of nuclear DNA)	1000 (500 on each nucleolar chromosome)	2 (in some cells two nucleoli fuse to form one)
Oocyte (after amplification)	37	4	12	25 (i.e 1000 times more than expected of a tetraploid cell; this would require 10 rounds of compound replication if all chromosomal gene sets were used)	2×10^6	1000–1500 (each containing 2–4 replicas of the rDNA gene complex)

All figures are subject to a 10 per cent variation according to author. These figures are based primarily on those of Dawid *et al.* (1970), Brown (1967), Birnstiel *et al.* (1971); and Macgregor (1972).

[†] Ribosomal DNA (or rDNA) is used here to include those sequences in chromosomal DNA which code for 18s and 28s ribosomal RNA, as well as the non-transcribed spacer regions; see fig. 3 for definition of 'gene set'.

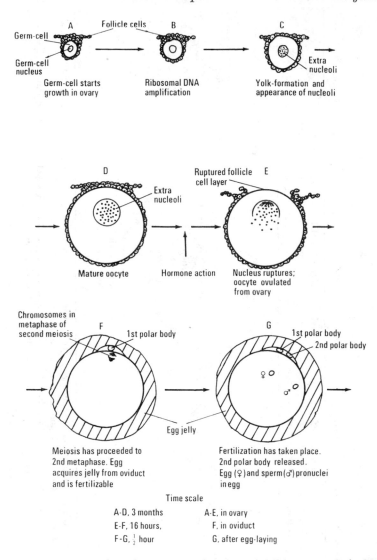

FIG. 3 Diagram of nuclear events during amphibian oogenesis (egg formation), meiotic maturation, and fertilization. Under natural conditions, hormones secreted by the pituitary gland cause maturation of fully mature oocytes, when their concentration in the blood reaches a certain level (see Smith and Ecker, 1970, for a review) Polar bodies are nuclei which are eliminated from the oocyte after its first and second meiotic divisions.

bands (Crouse and Keyl 1968; Pavan and da Cunha 1969). The localized increase in DNA is much more than two-fold and cannot therefore be attributed to out-of-phase early or late replication. The biological significance of this phenomenon is not clear because it occurs in larvae after the salivary gland cells have differentiated and shortly before they are resorbed at the time of pupation; the synthesized DNA does not necessarily have a coding function, and the same phenomenon certainly does not take place in polytene salivary gland chromosomes of all Dipteran species.

Does gene amplification take place commonly in development? It seems clear that it does not, as is evident from the following examples, which make use of molecular hybridization. This is a technique by which complementary strands of RNA or DNA are matched together *in vitro*; it can be used to determine the number of DNA sequences in a genome which are complementary to a certain kind of RNA or gene-sized fragment of DNA. The purification of mRNA for a known gene enables molecular hybridization experiments to be carried out with sufficient precision to recognize only a few genes of one kind in a genome. In the silk moth, the proportion of total DNA complementary to silk fibroin mRNA is very small, probably that needed for only 1–3 genes per haploid set; the amount of fibroin DNA is the same in the posterior silk gland (which is intensely active in fibroin synthesis) as it is in the middle silk gland and other larval cells, where fibroin is not synthesized to a detectable extent (Suzuki *et al.* 1972). Using highly radioactive DNA complementary to globin mRNA, Packman *et al.* (1972) have concluded from the kinetics of reassociation of globin genes in DNA extracted from duck erythrocytes and liver, that both these tissues have the same small number (<5) of globin genes. A similar design of experiment has been carried out with precision in respect of chick ovalbumin genes, which are expressed fully in oviduct cells. Sullivan and his co-workers (1973) conclude that ovalbumin genes are present at the same abundance in liver as in oviduct tissue, and that there is only one ovalbumin gene per haploid genome. All these three examples involve measurements of the numbers of copies of a gene whose product constitutes over 50 per cent of the total protein made by a cell. Since selective gene amplification does not take place in these cases, it is very doubtful if it occurs in respect of any other protein-coding genes.

Even ribosomal gene amplification, which is observed during oogenesis in some species, is not universal. It does not take place, for example, in some insect species whose oocyte ribosomal RNA is synthesized primarily by nurse cells (examples cited by Gall 1969). In spite of a number of investigations, ribosomal DNA amplification has not been observed in those somatic tissues where there is a rapid increase in the rate of ribosomal RNA synthesis (Brown and Weber, 1968). During *Xenopus* oogenesis only the region of DNA containing the 28s and 18s ribosomal genes undergoes amplification, and the 5s RNA genes do not (Brown and Dawid 1968). This argues against the possibility that gene amplification applies to all genes whose final product is RNA and not protein. The correct conclusion regarding localized increases in gene number is clearly that a mechanism exists in some cells by which gene amplification *can* take place, but that it is used only exceptionally, and is not a common process in cell differentiation.

There remains the possibility that the genome undergoes qualitative changes, involving the loss or change of single genes; this possibility has been extensively investigated in different ways which are now discussed in detail.

1.2. Changes in the informational content of individual genes

1.2.1. *Biochemical and cytological analysis*

The precision of molecular hybridization experiments may soon reach the point at which they can be used to distinguish between the presence or absence of a gene sequence in only one copy per genome in a sample of cells. However, most animal tissues consist of mixtures of several different cell-types, all of which will be represented in a sample of DNA used for hybridization. Consequently these methods cannot in general give information about a single cell-type. *In situ* hybridization (Gall and Pardue 1969; Jones 1970) permits the recognition of DNA sequences in cells of known type, but even when RNA or DNA of the highest specific activity is used, the limited efficiency of autoradiography restricts the use of this technique to situations where about 100 average-sized genes are located close to each other. There is, however, a more general limitation affecting the usefulness of molecular hybridization techniques for detecting qualitative changes in the genome of somatic cells. The composition of only a few base pairs could determine whether a gene is transcribed or not, and it is unlikely that it will be possible to recognize changes on such a small scale by molecular hybridization methods in the near future. On the other hand such small changes in the genome would be detected by the methods discussed below where a single somatic cell or nucleus is propagated in such a way as to promote the development of a normal organism. This is so because although the proteins synthesized in development may reflect the composition of only a small fraction of the genome's total DNA, any developmentally significant change to the rest of the genome (for example to regulatory genes) should result in abnormal development.

The cytological study of animal embryos has shown that in some invertebrate species, primarily Dipteran and Hemipteran insects, a substantial number of chromosomes are lost from nuclei during certain early cleavage divisions. The loss, which can affect the majority of chromosomes (e.g. 32 out of about 40 at the fifth cleavage division in *Mayetiola;* Bantock 1970), results in only the germ-line cells retaining the full chromosome complement, and all somatic cells having the reduced number. Chromosome elimination is therefore a type of genomic change which does not account for differences between somatic cells, but which could promote somatic versus germ-line differentiation. It is known that if elimination of chromosomes is induced experimentally in cells that would have become germ-cells, then germ-cells are not formed and sterile, but otherwise normal, individuals are obtained (Geyer-Duszyńska 1959; Bantock 1970). On the other hand it has not been possible to inhibit experimentally the elimination of chromosomes from somatic cells, nor therefore to provide a direct test of the possibility that somatic cells would differentiate just as well as if they had not undergone elimination. If nuclear DNA from cells with eliminated chromosomes (this is true of 90 per cent of *Ascaris* larval cells) is compared by hybridization with DNA from non-eliminated cleavage stages, the DNA which is eliminated appears to contain sequences not present in the rest of the genome (Tobler, Smith, and Ursprung 1972). This result, as well as cytological analysis (Niklas 1959), shows that the eliminated DNA is not the same as that which is retained, but it is still not certain that the eliminated DNA contains transcribable sequences, nor that the loss of chromosomes is causally connected with somatic cell differentiation.

In conclusion, direct methods of comparing the DNA of different cell-types at present lack sufficient sensitivity to reveal changes affecting genes present in one copy per genome. Major chromosome losses are sometimes associated with somatic versus germ-line differentiation but no such phenomenon is detectable in the vast majority of animal species.

1.2.2. *The growth of different cell-types from a single somatic cell*

The principle of the developmental method of comparing the genomes of different cells is to propagate mitotically the genome of a single somatic cell, and then demonstrate the capacity for different types of differentiation among the mitotic progeny of that cell. This approach has the great advantage of immense sensitivity in its ability to demonstrate the *presence* of genes for specified functions. It is not yet possible by other means to demonstrate the existence of single unexpressed genes in a single cell. But if a cell can be made to divide many times and so form an embryo or organism containing many different cell-types, then it is clear that that cell must have contained genes for each new type of protein synthesized.

In several plant species it has been possible to achieve the spectacular result of growing a mature plant from a single somatic cell. Though primarily concerned with animal development, we now describe these experiments on plants, because no comparable results have been obtained in any multicellular animal. Over many years, Steward and his colleagues (review by Steward 1970) have developed methods of stimulating the growth and differentiation of isolated carrot cells (*Daucus carota*) (Fig. 4). A 2–3 mm cube of carrot root phloem is cut out and incubated in a shaking culture as a result of which free cells are released. The dispersed cells are subjected to a series of conditions which include varying amounts of hormones, inbalanced nitrogen supply, more or less growth factors from coconut milk, low and high osmotic pressure, etc. After many weeks of such treatments, a mature carrot plant with normal flowers and seeds, and all somatic cell-types, can be grown. In another series of experiments, Vasil and Hildebrandt (1965), working with tobacco stem pith, grew pieces of excised pith on agar for several weeks, and obtained free cells from the resulting callus tissue by shaking it in a dissociating medium. An important feature of their procedure is that they isolated single cells

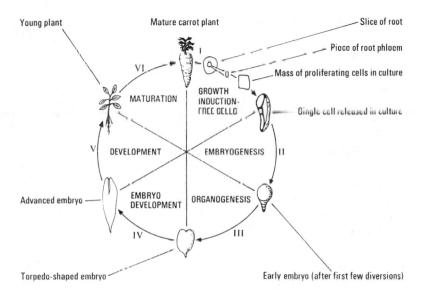

FIG. 4 Totipotency of carrot root cells, and environmental conditions which promote growth and development. This diagram shows the stages by which an isolated root cell can be grown into a complete plant. At each stage, the extent and normality of growth and development can be influenced by changing the components of the environment in the following ways: I. *Growth and proliferation*: promoted by factors present in coconut milk, such as cytokinins and indole acetic acid. II. *Embryogenesis*: promoted by coconut milk and napthalene acetic acid, with subsequent removal of auxins from basal medium to promote later development. III. *Organogenesis*: promoted by plentiful, balanced nitrogen supply and a low level of auxins. IV. *Further embryonic development*: promoted by high nitrogen supply and high osmotic pressure. V. *Plant development*: promoted by low osmotic pressure. VI. *Maturation*: with formation of storage root. (From Steward 1970.)

by hand, and about 30 per cent of such cells were stimulated to grow and divide. By use of a series of different media and conditions, they were able to obtain normal flowering tobacco plants from these single cells.

When interpreting these results, it is important to distinguish between somatic cells and differentiated cells. A substantial proportion, and probably a majority, of the cells contained in root phloem and stem pith are described as parenchymal, and almost certainly this is the type of cell from which whole carrot and tobacco plants have been grown. Cells of this kind are somatic in that they would never normally yield progeny which become gametes. However, unlike specialized animal cells, plant cells do not in general contain large amounts of cell-type specific proteins, and it is hard to characterize the differentiated state of an individual plant cell in terms of the proteins which it synthesizes. A characteristic which reflects plant cell differentiation is the accumulation of cell-type specific carbohydrates in the cell wall, but a parenchymal cell is not obviously differentiated in this or in other respects. The experiments cited above therefore demonstrate that the loss or permanent alteration of genes cannot be a general characteristic of plant somatic cells. To the limited extent that a parenchymal cell may be regarded as representative of specialized cells, this conclusion is also valid for differentiated plant cells.

In animal cells, it has proved conspicuously hard to induce any change in cell-type in somatic cells. The stability of the differentiated state (Ursprung 1968) has been established by grafting cells to an abnormal position in the body, as well as by culturing cells *in vitro*. The best example of the first kind is the remarkable persistence of the determined state in serial grafts of imaginal discs in *Drosophila* (Hadorn 1968). The stability of differentiation in cultured cells is exemplified by cartilage cells which can maintain their differentiated state for well over 20 cell generations under appropriate conditions (Coon 1966; Holtzer and Abbott 1968).

Specialized animal cells can change rather easily from an overtly differentiated state, in which they synthesize cell-type specific proteins, to what is sometimes called a dedifferentiated condition, when the cells appear morphologically and bio-chemically undifferentiated. The important point is that dedifferentiated cartilage or skin cells, or their mitotic descendants, will never form another specialized cell type, such as a muscle or blood cell; all they can do is to fluctuate between an overtly differentiated and a 'determined' state. The inability of animal cells to change from one specialized type to another is a generalization to which there appears to be only one kind of certain exception. This is the fact that the dorsal pigmented iris cells of the eye of certain urodele amphibia (and of a few) other vertebrates can divide and transform into lens cells containing normal lens crystallins; this process is known as Wolffian regeneration and has been reviewed by Reyer (1962) and Yamada (1967a). In some species a lens is said to be regenerated from corneal or retinal cells (Clayton 1970; Eguchi and Okada 1973). Except in these cases, the transformation of one cell-type into another rarely if ever occurs in animals; evidently there is in animal cells a special mechanism which stabilizes and propagates the determined state. However, Wolffian regeneration suggests that there can be no underlying reason why such transformations are impossible once the normal control processes are overcome. It should be added that many animal species contain cells, such as fibroblasts and mesenchyme cells, which, like plant parenchymal cells, do not obviously conform to any one type of specialization. These may give rise to daughter cells which differentiate in various recognizable ways. Nevertheless, such examples do not prove that one cell-type can transform into another because the apparently unspecialized cells may be a heterogeneous class containing cells already determined in different ways.

To demonstrate conclusively the existence of unexpressed genes in specialized or determined animal cells, it is necessary to transfer a nucleus into a new cytoplasmic environment, thereby releasing its genes from the controls that normally operate in specialized cells. Since nuclear-transfer experiments have provided what is at present the best evidence for the qualitative constancy of the genome in all cells of an individual, they are now discussed in detail.

1.3. Nuclear transfer experiments—evidence for the genetic identity of somatic cell nuclei

1.3.1. *Nuclear transfer technique*

Nuclear transplantation is a method whereby the living nucleus of one cell is introduced into the cytoplasm of another cell whose own nucleus has been removed or destroyed. The method was first carried out successfully in *Amoeba* by Commandon and de Fonbrune (1939). Since then the technique has been applied to a number of problems connected with nucleocytoplasmic interactions in Protozoa, for example by Danielli *et al.* (1955) and Goldstein and Prescott (1967) in *Amoeba*, and by De Terra (1969) in the ciliate *Stentor*. Some early experiments have been reviewed by Gurdon (1964).

It has been obvious to embryologists for a long time that it would be very desirable to be able to transplant a somatic cell nucleus to the enucleated egg of an animal, thereby providing a direct test of whether the genes of the somatic cell nucleus are able to substitute for the fertilized egg (zygote) nucleus in promoting normal development. The first real success in transplanting a living nucleus in animal cells was achieved by Briggs and King (1952) in the frog *Rana pipiens*. Since then, nuclear transplantation has been successfully applied to many amphibian species, and to *Drosophila*, but to no other vertebrates and so far with only limited success to some other insect species (including the bee *Apis*—Dupraw 1967, and a beetle *Leptinotarsa*—Schnetter 1967). Work on Amphibia has provided a direct test of whether somatic cell nuclei are developmentally equivalent to a zygote nucleus. All amphibian nuclear-transplant experiments fall into two categories—those which demonstrate the genetic similarity of all somatic cell nuclei, and those which are concerned with differences between somatic and zygote nuclei. The most convinɣing evidence against the loss or stable alteration of genes in somatic cells is provided by the normal development of enucleated eggs injected with the nuclei of specialized cells from the frog *Xenopus laevis*.

The technique of transplanting nuclei in Amphibia is shown diagrammatically in Fig. 5. It entails three steps. The isolation of donor cells (step 1) and enucleation of recipient eggs (step 2) are not technically difficult. The third step involves sucking up

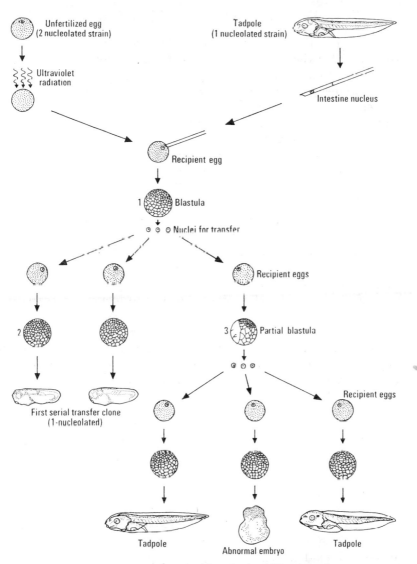

Unfertilized egg
(2 nucleolated strain)

Ultraviolet
radiation

Tadpole
(1 nucleolated strain)

Intestine nucleus

Recipient egg

1 Blastula

Nuclei for transfer

Recipient eggs

First serial transfer clone
(1-nucleolated)

2

3 Partial blastula

Recipient eggs

Tadpole

Abnormal embryo

Tadpole

Second serial transfer clone (all 1-nucleolated)

Fig. 5 Nuclear transplantation procedure for Amphibia. The diagram shows recipient egg enucleation by u.v. light, first nuclear transfers, and serial nuclear transfers, as used in *Xenopus*. In this Figure, a first-transfer blastula (1) is used to make serial transfers. Two of the complete serial-transfer blastulae (2) are allowed to develop into tadpoles. A third serial-transfer blastula (3) is only partially cleaved; it is used to prepare a second serial-transfer clone, which includes normal as well as abnormal tadpoles and embryos in it. (From Gurdon 1968*a*.)

an isolated cell into a pipette whose internal diameter is just too small for the cell, but is large enough to avoid removing cytoplasm from around the nucleus. As a result the broken cell with its nucleus still surrounded by cytoplasm is injected into the enucleated egg. This part of the technique requires substantial practice and skill if optimal results are to be obtained. The injection of a very small amount of donor cell cytoplasm, at the same time as the nucleus, does not affect the conclusions drawn below from these experiments.

1.3.2. *Specialized cells used to provide donor nuclei*

Two series of experiments have given the most decisive results. The first was carried out with nuclei from intestinal epithelium cells of feeding tadpoles (Gurdon 1962a). This cell-type was chosen as a nuclear donor because its differentiated state is clearly evident from its possession of a striated border, a specialization related to its absorptive function. Although the intestine of a swimming tadpole contains muscle and other non-epithelial cells, the epithelial cells can be easily distin guished from these. They are much larger, they are the only larval intestine cells which have not resorbed all their yolk, they dissociate rapidly in a low concentration of versene, and they can often be seen to possess a 'brush border' when dissociated. Compared to other specialized cells, tadpole intestine cells divide unusually rapidly (in connection with gut extension), and this probably accounts for why their nuclei can be transplanted more successfully than other specialized tadpole cells.

For the second series of experiments, nuclei were taken from the cells of adult frog tissues, such as lung, kidney, skin, etc. (Laskey and Gurdon 1970). Satisfactory development of embryos can be obtained from the nuclei of adult organs, but only if explants are made, and donor nuclei taken from the cultured cells which grow out within a few days. Very poor results are obtained if nuclei from adult cells are transplanted immediately after cell dissociation, rather than after culture for a few days (King and DiBerardino 1965; Gurdon and Laskey 1970). This is possibly due to the fact that such cells divide less rapidly than cells grown in culture. The cultured cells which grow out from an explant are not normally identifiable as specialized cells of the donor tissue (such as lung, kidney or liver cells) and may be 'fibroblasts' contained within that tissue. This is not the case, however, with adult skin; the cells which grow out from a small piece of adult frog skin (dermis and epidermis) can be identified as committed skin cells by the following criteria: (1) Between 6 and 8 days after explantation, all cells which grow out fill up spontaneously with a highly birifringent material, resembling keratin; this process is never observed in cells grown out from any other tissue, such as lung, kidney, heart muscle, testis. (2) The acquisition of birifringent material observed in the cells that grow out from skin tissue is inhibited by Vitamin A, and accelerated by citral, an analogue of Vitamin A. This effect has been described in detail by Fell and co-workers (e.g. Aydelotte 1963) for cultures of chick skin; it is peculiar to, and is often regarded as diagnostic of, keratinization. (3) Recently Reeves (unpublished results) has been able to prepare antibodies to a keratin like material extracted from adult frog skin. When made fluorescent, these antibodies bind to over 99 per cent of all cells which grow out from an explant of foot web, but do not bind to cells grown out from explants of other organs or to cultured frog fibro-blasts. The cells grown out from skin bind the anti-skin anti-bodies 5 days after explants are made. Successful nuclear transfers have been performed with cells 6–7 days after explant-ation. This is about 24 hours before they would have acquired visible amounts of keratin-like material, and would therefore have become impossible to handle for nuclear transplant-ation.

A very important characteristic of the intestine and skin cells used for these experiments is their possession of a genetic nuclear marker. An almost ideal nuclear marker for nuclear transfer experiments exists in the form of the 'Oxford' O-*nu* mutant discovered by Fischberg (Elsdale *et al.* 1958); its characteristics are shown in Fig. 6. In nuclear transfer experi-ments, donor nuclei taken from 1-*nu* cells heterozygous for the mutant were transplanted to recipient eggs of wild-type animals (Elsdale *et al.* 1960). Each nuclear-transplant embryo was then proved to contain nuclei derived from the trans-planted nucleus; failure to eliminate the egg-nucleus, which is killed by U.V. irradiation in *Xenopus* (Gurdon 1960*a*, and

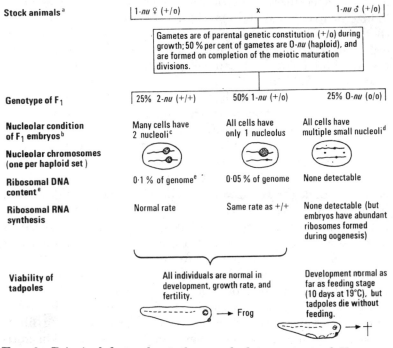

Stock animals[a]

1-*nu* ♀ (+/o)   x   1-*nu* ♂ (+/o)

Gametes are of parental genetic constitution (+/o) during growth; 50 % per cent of gametes are 0-*nu* (haploid), and are formed on completion of the meiotic maturation divisions.

Genotype of F$_1$

25% 2-*nu* (+/+)   50% 1-*nu* (+/o)   25% 0-*nu* (o/o)

Nucleolar condition of F$_1$ embryos[b]

Many cells have 2 nucleoli[c]   All cells have only 1 nucleolus   All cells have multiple small nucleoli[d]

Nucleolar chromosomes (one per haploid set)

Ribosomal DNA content[e]

0·1 % of genome[e]   0·05 % of genome   None detectable

Ribosomal RNA synthesis

Normal rate   Same rate as +/+   None detectable (but embryos have abundant ribosomes formed during oogenesis)

Viability of tadpoles

All individuals are normal in development, growth rate, and fertility.
→ Frog

Development normal as far as feeding stage (10 days at 19°C), but tadpoles die without feeding.
→ +

FIG. 6 Principal facts about the anucleolate mutant of *Xenopus laevis*. The stock is maintained as heterozygotes, whose nuclei have one chromosome with the normal number of ribosomal gene sets (and hence one nucleolus), and a homologous chromosome with no ribosomal genes and no nucleolus. The diagram starts with a mating between a male and female, both heterozygous for the mutation.

(a) The *Xenopus* 0-*nu* mutant was first described by Elsdale *et al.* (1958); the absence of rRNA synthesis in 0-*nu* by Brown and Gurdon (1964); the deficiency of ribosomal DNA by Wallace and Birnstiel (1966); partial nucleolar mutants by Miller and Gurdon (1970), Knowland and Miller (1970) and Miller and Knowland (1972). The genetics of *X. laevis* has been summarized by Gurdon and Woodland (1974). (b) The larval tail-tip is cut off and subsequently regenerated. It can therefore be used for phase contrast examination or for making whole mounts, without preventing the identified larvae from growing to normal adults. (c) In embryos, 70 per cent of all cells have two nucleoli, and 30 per cent have only one nucleolus. In some specialized adult organs only 12 per cent have two nucleoli (Wallace 1963). (d) In some tissues of late larvae, the largest 'pseudonucleoli' resemble normal nucleoli, but 0-*nu* nuclei are easily recognisable because they also contain *several* extra small nucleoli. (e) 0·01 per cent is the proportion of the genome complementary to 28s+18s RNA. The ribosomal cistrons including the spacer region amount to about 0·2 per cent of the genome (see Table 1 and Fig. 2).

$^+/_+$, wild-type; $^+/_0$, heterozygous for the mutation; $^0/_0$, homozygous mutant.

TABLE 2

Proof that the nuclei of a nuclear-transplant embryo are derived from the transplanted nucleus and not from the egg nucleus†

Material tested	Nature of test	Purpose of test
Female frog which laid recipient eggs	Cells grown from skin, and explants tested for number of nucleoli and number of chromosomes	Recipient eggs laid by frog proved to be genetically 2-*nu* (wild-type) and diploid
Donor cells cultured from adult organ explant	Determination of numbers of nucleoli and chromosomes per cell	Donor cells proved to be 1-*nu* and diploid
Nuclear-transplant tadpoles from 1-*nu* diploid nuclei transplanted to 2-*nu* wild-type eggs	Phase-contrast study of tail-tip squash preparations	Nuclear-transplant tadpoles contain only 1-*nu* cells
Nuclear-transplant tadpoles from 1-*nu* diploid nuclei transplanted to 2-*nu* wild-type eggs	Colchicine treatment of tissue from body of tadpole	All chromosome preparations found to be diploid (2N±2)
Nuclear-transplant tadpoles from 1-*nu* diploid nuclei transplanted to 2-*nu* wild-type eggs	Tail-tip of tadpole fixed, and diameters of flattened epidermal nuclei measured	At least 99 per cent of all cells from tail tissue had nuclei of diploid diameter (clearly distinguishable from haploid, triploid or tetraploid nuclei). Therefore mosaic constitution excluded.
Nuclear-transplant tadpoles from 2-*nu* diploid donor nuclei transplanted to 2-*nu* wild-type eggs‡	Living and fixed cells from the resulting tadpoles examined for numbers of nucleoli and chromosomes	Each tadpole contained diploid cells, most of which were 2-*nu*. Therefore enucleation does not convert 2-*nu* nuclear marker to 1-*nu* condition.‡

† This analysis was carried out on nuclear-transplant tadpoles prepared from adult skin cell nuclei (from Gurdon and Laskey 1970).

‡ Cultured cells which are genetically 2-*nu*, or wild-type, mostly contain only 1-nucleolus (resulting from fusion of two nucleoli). In this experiment many *phenotypically* 1-*nu*, but genetically 2-*nu*, nuclei were transplanted, and always gave nuclear-transplant embryos containing the expected proportion of nuclei with two nucleoli. This proves that the *phenotypic* 1-*nu* condition is reversible, and not mitotically stable like the repressed X-chromosome of mammalian cells.

Fig. 5), was thereby excluded. In the cultured skin cell experiments, exhaustive tests were performed to eliminate the presence in a transplant-embryo of even a few nuclei derived from the egg nucleus; even the possibility that U.V. irradiation might change the genetically 2-*nu* condition of the egg nucleus into a 1-*nu* condition was excluded (Table 2). The fact that every nuclear-transplant embryo was individually proved to be wholly of donor nucleus origin cannot be overemphasized.

1.3.3. *Usefulness of serial transfers*

When transplanting nuclei from specialized cells, results are much improved if use is made of serial nuclear transfers (Fig. 5). For this purpose, a recipient egg which has received an intestine or skin cell nucleus is grown to the blastula stage (*ca.* 10 000 cells); the cells of this first-transfer blastula are dissociated and used as nuclear donors for another (serial) transfer of nuclei to a further set of recipient eggs. Rather surprisingly, the development of serial-transplant embryos is much improved if nuclei are taken from partially cleaved, as opposed to complete, first-transfer embryos (Gurdon 1962*a*; Gurdon and Laskey 1970). Partial blastulae always die before the completion of gastrulation, but some of the serial transplant-embryos prepared from nuclei of a partial blastula develop further than this and therefore much more normally than the first-transfer embryo from whose nuclei they are derived. The explanation for this effect which is not observed in subsequent serial-transfer clones is discussed later under the development of abnormal nuclear-transplant embryos (p. 32). For nearly all experiments with intestine and skin cell nuclei, use has been made of serial transfers from partial blastulae (Gurdon 1962*a*; Gurdon and Laskey 1970).

1.3.4. *Results with* Xenopus

A complete test for the presence of all genes in a transplanted nucleus would require that some of the first or serial transfer embryos derived from it should develop into normal and fertile adult frogs. Fertile male and female adults have been prepared from transplanted nuclei of early *Xenopus* embryos (Gurdon 1962*b*). Furthermore it has been possible to prepare normal fertile males and females from a small proportion of

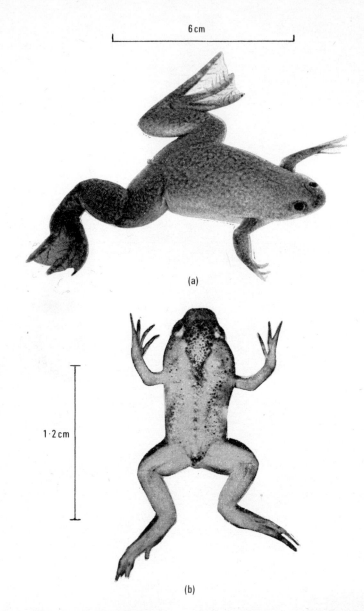

6 cm

(a)

1·2 cm

(b)

FIG. 7 Frogs produced by the transplantation of nuclei from cultured cells of *Xenopus laevis*. (a) A normal adult frog resulting from the serial transplantation of nuclei from cells which had been grown in culture for over a month. The cell culture was derived from a piece of epidermis of a swimming tadpole. (b) A recently metamorphosed frog which resulted from the serial transplantation of nuclei of cultured skin cells. The cultured cells were used as nuclear donors one week after they had grown out from a small piece of foot-web skin of an *adult* frog.

nuclei taken from intestinal epithelium cells of feeding tadpoles (stage 46) (Gurdon and Uehlinger 1966; Gurdon 1968a), and from swimming tadpole (stage 40) epidermis (Brun and Kobel 1972). The most normal development yet achieved from an adult skin cell nucleus is a young frog which died a few weeks after metamorphosis (Fig. 7b), up to which stage it appeared normal.

Fischberg *et al.* (1963) have carried out a genetic analysis of adult nuclear-transplant frogs, to find out whether recessive mutations arise in the course of cell differentiation. A nuclear-transplant frog was mated with the father or mother of the tadpole from one of whose nuclei it was derived. The resulting F_1 frogs were mated among themselves; the presence of a recessive mutation in the transplanted somatic cell nucleus was revealed by developmental abnormalities in the F_2 progeny, in one of four such matings. If the mutation was already present in the stock, this was revealed by the progeny of backcrosses between the nuclear-transplant frog, or its progeny, and the parent frogs of the transplanted nucleus. The result of this analysis is that recessive mutations were found as often in the wild-type stock animals as they were in nuclear-transplant embryos (Fischberg *et al.* 1963). There is therefore no reason to think that such mutations arise more commonly during somatic cell differentiation than they do in germ-line cells. This type of analysis is very time-consuming, requiring 3–4 years for the analysis of a single somatic cell nucleus. An accelerated procedure for the genetic analysis of a somatic cell is described in Appendix B.

The general conclusion to be drawn from a study of adult nuclear-transplant frogs is that at least some intestine nuclei carry an entirely normal set of genes, including those concerned with gametogenesis. However the frequency with which fertile adults are obtained from tadpole intestine nuclei is small (1–2 per cent), and the case for believing a complete set of genes to be present in the nuclei of specialized cells is greatly strengthened by the fact that a higher percentage of nuclear-transplant embryos reach the normal larva stage.

The results of tadpole intestinal epithelium and adult skin cell nuclear-transplants are summarized in Table 3. In both experimental series progressively fewer embryos survive to more advanced developmental stages. The most meaningful way to express nuclear transplantation results is in terms of the percentage of nuclear-transplant eggs which reach developmental stages with specialized cell-types of a well-defined kind, i.e. those which synthesize cell-type specific proteins. We are primarily interested in the proportion of transplanted nuclei which can 'form' young tadpoles with functional muscle, nerve, lens, and blood cells, and therefore in the proportion of nuclei which, having already been predominantly active in intestine or skin cell differentiation, have nevertheless retained functional genes for myosin, nerve cell proteins, crystallin, and haemoglobin. As is seen in Table 3, about 20 per cent of intestine nuclei, and 12 per cent of cultured skin nuclei, can promote

TABLE 3

The development and differentiation of embryos prepared by transplanting nuclei from specialized cells into enucleated unfertilized eggs of Xenopus laevis

| | | Per cent of total transfers reaching: | | | |
| | | Tadpoles with functional muscle and nerve cells | | Tadpoles with normal muscle, nerve, lens, heart, blood, etc., cells | |
Donor cells	Total transfers	1st transfers	1st and serial transfers	1st transfers	1st and serial transfers
Intestinal epithelial cells of feeding tadpoles†	726	$2\frac{1}{2}$	20	$1\frac{1}{2}$	7
Cells grown from adult frog skin‡	3546	0·1	12	0·03	8
Blastula or gastrula endoderm‡	279	48	65	36	57

† from Gurdon (1962a).
‡ from Gurdon and Laskey (1970).

the formation of nerve and muscle cells, and 7–8 per cent of these nuclei support tadpole development as far as a stage which includes lens, blood, etc., cells as well. Recently, Kobel *et al.* (1973) have shown that just under 1 per cent of the nuclear transfers made from cultured melanophore cells of stage 35–40 tadpoles were able to promote development as far as a hatching tadpole (stage 30, with functional muscle and nerve cells). The important point demonstrated by these results is that the mitotic products of a significant proportion of the transplanted intestine and skin cell nuclei can support the differentiation of totally different cell-types. The success rate obtained with intestine nuclei is high enough to exclude the possibility that the normal development is attributable to unknown cell-types contaminating the donor cell preparations. In the case of adult skin cells, over 99 per cent can be shown to contain keratin. These experiments therefore demonstrate that intestine and skin cells have not undergone a loss or stable inactivation of genes for haemoglobin, myosin, lens crystallins, etc. This is the basis of the generalization that differential gene expression during animal development does not entail the loss, stable alteration, or permanent inactivation of genes which are never normally expressed during the life of a somatic cell and its mitotic progeny.

1.3.5. *Results with other species*

Although the nuclear transfer results obtained with *Xenopus* have given more normal development than is obtained with other species, they are supported by experiments on other amphibian species. There is a wide spread of success rate in amphibian nuclear-transfer experiments, ranging from the Axolotl which gives very little normal development from nuclei of postneurula embryos (Signoret *et al.* 1962; Briggs *et al.* 1964), through *Rana pipiens* (Briggs and King 1957), to *Pleurodeles* (Picheral 1962), and *Bufo bufo* (Nikitina 1964) which gives results nearly as good as those obtained with *Xenopus*. Even in *Rana pipiens*, it has now been possible to prepare a nearly normal tadpole (with all major cell types) from the kidney or adenocarcinoma cells of a newly metamorphosed frog (McKinnell *et al.* 1969), and normal larvae from lens epithelial cells (Muggleton-Harris and Pezzella 1972). Furthermore nuclei from the endoderm cells of *Rana pipiens* embryos give nuclear transfer results as good as those of *Xenopus* if the donor cells are treated with the polyamine spermine before nuclear transfer (Hennen 1970). In *Rana* there is substantial variation in the extent to which different investigators obtain normal development from the same kind of nuclei (Fig. 9).

Apart from Amphibian species, *Drosophila* is the only animal in which the developmental capacity of the nuclei of differentiating cells has been tested; attempts to transplant nuclei in mammals are referred to in Appendix A. Following the first successful transfers of multiple cleavage nuclei in

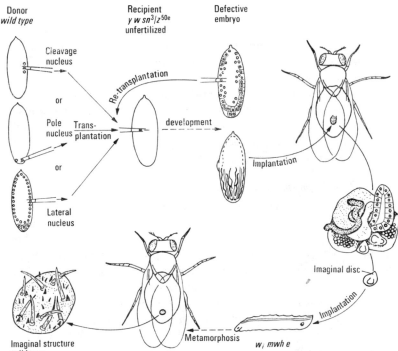

FIG. 8 Nuclear transplantation in Drosophila. This diagram shows the principal steps involved in testing the developmental capacity of *Drosophila* blastoderm nuclei. These include first nuclear transfers, serial nuclear transfers, imaginal disc transfers to an adult haemocoel (for proliferation of cells), and imaginal disc transfers to larvae (so that the discs differentiate into adult structures in the course of development through the pupal stage). The small letters in italics indicate genetic mutants used to distinguish host and donor cells. From Illmensee (1972). See Hadorn (1968) for explanation of imaginal disc transfers.

Drosophila (Geyer-Duszyńska 1967; Illmensee 1968), Schubiger and Schneiderman (1971) and Zalokar (1971) obtained adult cell-types and gametes, carrying donor nucleus markers, from fertile eggs which were injected with nuclei and which developed as nuclear mosaics. The successful transplantation of single nuclei from a blastoderm embryo into unfertilized eggs was first achieved by Illmensee (1972). Two to four per cent of nuclei from three different regions of an early embryo promoted development to the hatching larva stage, and in a few cases to mid-larval stages. By growing abnormal embryos in an adult abdominal cavity, and then transferring the imaginal discs obtained in this way to larvae, it was shown that blastoderm nuclei are capable of promoting the differentiation of adult structures such as wing, halteres, and antennae (Fig. 8). Although no differences were observed in the developmental capacity of nuclei from different regions of cleavage or blastoderm embryos, it has not yet been possible to test nuclei from imaginal disc cells which are certainly 'determined' for a limited range of differentiations.

In conclusion, there is no fundamental disagreement between the results obtained in *Xenopus* and those obtained from nuclear transfer experiments on any insects or amphibian species; nuclear transfer experiments on all species are consistent with the view that specialized cells contain a complete range of genes, most of which are not expressed in any one specialized cell-type.

1.4. Nuclear transfer experiments: loss of developmental capacity in transplanted nuclei

Almost as soon as the technique of transplanting nuclei in amphibia was worked out, King and Briggs (1955) pointed out that their results suggested a progressive loss, during development, of the ability of nuclei to promote normal embryogenesis, when tested by nuclear transplantation. This at once raised the important possibility that embryonic development might be accompanied by the loss or permanent inactivation of genes. Now nearly 20 years later, these and all subsequent observations are best interpreted within the framework of four generalizations.

1.4.1. *The developmental capacity of transplanted nuclei decreases with increasing age of donor*

The first generalization is that nuclei from more advanced developmental stages and from more differentiated cells always promote less normal nuclear-transplant embryo development (quantitatively and qualitatively) than nuclei from early developmental stages or undifferentiated cells. This effect is observed, not only when different developmental stages are compared (Fig. 9), but also when a comparison is made between nuclei from more and less differentiated regions of the same tissue (Briggs *et al.* 1964). The decline in developmental capacity of transplanted nuclei has been observed in all species so far tested, though there is some variation in the pattern of decline in different species (Fig. 9).

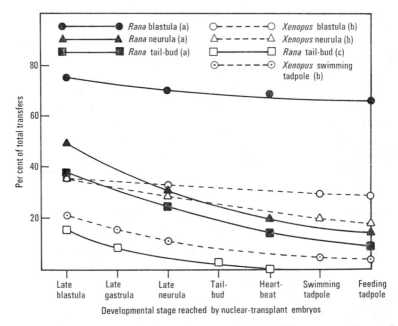

FIG. 9 Nuclear-transplant embryo survival. The proportion of nuclear-transplant eggs which develop normally declines as donor nuclei are taken from progressively more advanced stages. The results shown are for endoderm nuclei. Stages of normal development are illustrated in Fig. 29. (a) Hennen (1970); (b) Gurdon (1960b); (c) Briggs and King (1957).

A decline in developmental capacity does not necessarily reveal a deficiency in the genetic content of transplanted nuclei. This is clearly demonstrated by the results of transplanting nuclei from germ-cells, which as mitotic parents of eggs and sperm, certainly have a complete genome. In *Rana pipiens*, Smith (1965) obtained normal development from germ-cell but not from endoderm nuclei of tadpoles. Yet in the same species, DiBerardino and King (1966) were unable to obtain development beyond the mid-gastrula stage, using nuclei from primary spermatocytes of an adult testis. Indeed more normal development was obtained with adult kidney nuclei than with spermatocyte nuclei (King and DiBerardino, 1965).

1.4.2. *The reduced developmental capacity of transplanted nuclei is irreversible*

The second generally accepted conclusion from these experiments is that the developmental abnormalities of nuclear-transplant embryos are often caused by an irreversible nuclear condition. This was first established by the serial nuclear-transplant experiments of King and Briggs (1956), and subsequently confirmed in other species (Gurdon 1960*b*). This conclusion is based on a comparison of the developmental abnormalities observed among serial-transplant embryos derived from a single originally transplanted nucleus (Fig. 10). There is considerable variation in the extent to which embryos within a clone (see glossary) develop normally. However, consistent differences in development are commonly observed between the various clones derived from different nuclei of the same original donor embryo (Fig. 10). These observations show that the nuclei derived by transplantation from one original somatic nucleus differ in an irreversible (presumably genetic) way from the nuclei derived from another somatic nucleus. Indeed, as further serial-transfer clones are prepared, the developmental abnormalities observed in one clone are propagated, often in more severe form, by the embryos in all clones derived directly from it.

1.4.3. *Chromosome abnormalities are the cause of reduced developmental capacity*

We now come to the key question of the basis of the

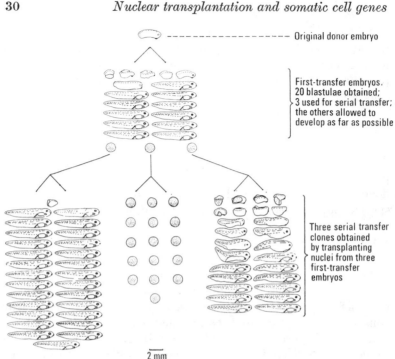

FIG. 10 Clonal variation in serial nuclear transfers. Endoderm nuclei from a post-neurula embryo promote both normal and abnormal development of nuclear-transplant eggs. Three normal first-transfer blastulae (whose expected development was not known at this stage) were taken to make three serial-transfer clones (explained in Fig. 5). 50 nuclear transfers were made in each clone, and this figure shows the furthest development of all eggs which reached the normal blastula stage. In one clone nearly all embryos developed into normal larvae, but in another all died at the late blastula stage. From Gurdon (1960*b*).

irreversible nuclear condition responsible for the abnormalities observed in nuclear-transplant embryos. It is now agreed that the great majority, and possibly all, of such developmental abnormalities are caused by chromosome abnormalities, which are themselves a consequence of nuclear transplantation; these are most severe in embryos which develop most abnormally (Table 4). This relationship was first observed by Briggs *et al.* (1960), and has since been substantiated in detail by several

TABLE 4

Relationship between abnormal development and abnormal chromosome constitution in nuclear-transplant embryos†

	Stage of developmental arrest			
	Blastula and gastrula	Neurula and tail-bud	Heart-beat and abnormal tadpoles	Controls from fertilized eggs (normal)
Percentage of mitoses with diploid ± 1 chromosome number	15 (often with loose chromosome fragments)	32	87	95
Percentage of mitoses with exactly diploid (26) chromosome number	0	18	62	88
Number of mitoses analysed	33	38	91	63
Number of embryos analysed	4	6	5	4

† These results were obtained from first and serial transfers of adenocarcinoma cell nuclei in *Rana pipiens*. Embryos were incubated in colchicine for several hours, so as to accumulate cells in mitosis; they were then squashed and stained in orcein to count chromosomes. The last two lines of the Table show that the conclusions are based on large numbers of chromosome counts. (From King and DiBerardino 1965).

other authors (e.g. Gallien *et al.* 1963; Hennen 1963). Chromosome abnormalities do not exist in the donor cells, and must therefore arise as a result of nuclear transplantation. This largely eliminates the possibility that the stable differences observed between serial clones may reveal normal genetic differences between somatic nuclei.

It is not known for certain why chromosome breakage occurs after nuclear transplantation, but it is very probably related to the fact that transplanted nuclei are often not completely replicated before the first cell division. Most somatic cells divide infrequently (every 1–2 days), and take about 7 hours to replicate their chromosomes; yet any transplanted nucleus which is to participate in normal development must start and finish replication of its chromosomes within one hour since eggs always undergo cytoplasmic cleavage within 1–2 hours of activation by micropipette penetration. The reasons for suggesting the above relationship are that chromosome abnormalities occur more often with nuclei from slow-dividing (differentiated) cells than with fast-dividing blastula nuclei, and that DNA synthesis in transplanted nuclei often extends beyond the normal time (Graham *et al.* 1966). Indeed, chromosomes can often be seen stretched between the two poles of a mitosis, presumably because the incomplete replication of chromosomes prevents them undergoing disjunction as usual (Briggs *et al.* 1964; Gurdon and Laskey 1970).

These facts suggest an explanation for the surprisingly beneficial effect, referred to above (p. 22), of carrying out serial transfers from partially cleaved blastulae. Such partial blastulae usually arise from eggs in which the transplanted nucleus fails to divide when the egg cytoplasm undergoes its first division to the two-cell stage. The effect of this is that the transplanted nucleus has twice as long as usual in which to complete chromosome replication, before dividing for the first time when the nucleated half of the egg undergoes division (Gurdon and Laskey 1970).

Chromosome abnormalities arise not only at the first but also at later divisions of transplanted nuclei (Briggs *et al.* 1964), and nuclear-transplant blastulae therefore consist of mosaics of cells with normal and abnormal nuclei. Probably because of the extra time available for the first chromosomal replication, the nucleated half of a partial blastula more often includes some normal nuclei than do the cells of the morphologically more normal complete transplant-blastulae.

1.4.4. *The question of cell-type specificity of nuclear changes revealed by transplantation*

The fact that nuclear-transplant embryo abnormalities are attributable to chromosome damage incurred as a result of nuclear transplantation, does not necessarily mean that they are without interest. It has been suggested (Briggs *et al.* 1964) that the regions of chromosomes which are 'late-replicating' and possibly transcriptionally inactive in somatic cells are also those which fail to replicate sufficiently quickly after nuclear transplantation. If for this or any other reason the chromosomal and developmental abnormalities in nuclear-transplant embryos were donor-cell specific (for example embryos from endoderm nuclei would be abnormal in respect of ectoderm- or mesoderm-derived structures) this would raise the possibility that these abnormalities might reflect differential gene expression in somatic cells. In fact, several karyotype analyses (unpublished) have failed to reveal *consistent* types of chromosomal differences between abnormal nuclear-transplant embryos derived from endoderm and ectoderm cells. Therefore if any consistent chromosomal abnormalities exist, they are obscured by other, apparently random, variations.

The fact that chromosomal abnormalities cannot be shown to be related to the origin of the donor nucleus does not exclude the possibility that nuclear transplant embryos with *normal* chromosome sets might display morphological deficiencies peculiar to the donor tissue of origin. A syndrome of abnormalities seen in nuclear-transplant embryos of endodermal origin has been described in detail (Briggs and King 1957), and involves the early death of ectodermal and mesodermal tissues (skin, nerve, muscle, etc.). However this does not necessarily prove that developmental abnormalities are donor-cell specific, because the tissues of endodermal origin (gut) differentiate later in development than do other tissues and it would not be surprising if endodermal tissues survived for longer in *all* kinds of abnormal larvae, whatever the cause of the abnormality.

For this reason the real test of donor-cell specificity requires not merely that endoderm tissues survive longest in endoderm embryos, but also that a *different* syndrome is observed in mesodermal or ectodermal embryos. The strongest claim to see such differences has been mounted by DiBerardino and King (1967), who believe that random chromosome abnormalities may obscure an underlying tendency for developmental abnormalities to be donor-cell specific. For this reason they discount all but 12 in a series of 200 nuclear-transplant embryos derived from nuclei of neural tissue (a part of the ectoderm), since these 12 were the only ones which had the diploid number of chromosomes. Of these, some had the right number but wrong shape of chromosomes, and only four had karyotypes uniformly indistinguishable from normal in all respects. Of these four, three contained mesodermal or endodermal abnormalities and one had deficiencies in cells derived from all three germ-layers. Therefore if any donor-cell specificity exists it is observable in only 2 per cent of all nuclear transplants and even among these, 25 per cent do not conform to the rule. Furthermore, histological examination of these tadpoles reveals that those described as suffering from deficiencies of the notochord and somites in fact contain cells cytologically recognizable as notochord and muscle cells, and it is only the abundance and arrangement of these cells which is abnormal. Since no other claims have been made for an ectodermal or mesodermal syndrome of abnormalities, it is clear that no strong case has yet been established for the existence of donor-cell specific abnormalities among nuclear-transplant embryos; this is the fourth generalization about abnormal nuclear transplant-embryo development.

1.5. Conclusions on the developmental capacity of transplanted nuclei

Nuclear transplantation experiments that concern genetic differences between somatic cells can be summarized as follows: (1) As development proceeds and cells become differentiated, their nuclei progressively lose their capacity to promote normal development of recipient eggs. (2) Many of the developmental abnormalities observed in nuclear-transplant embryos are caused by a nuclear condition which is stable as judged by its ability to be propagated mitotically in serial nuclear-transplant experiments. (3) The stable nuclear condition which promotes developmental abnormalities is due in the great majority, and possibly in all, cases to chromosomal abnormalities which arise soon after nuclear transplantation. (4) There is no convincing evidence that the chromosomal or developmental abnormalities of nuclear-transplant embryos are related in type to the kind of cells from which donor nuclei are taken. (5) The developmental and chromosomal abnormalities of nuclear-transplant embryos can be adequately accounted for if most slow-dividing somatic nuclei are unable to fully raise their rate of chromosome replication and division to the very fast pace set by egg cytoplasm; random chromosome breakage then occurs, when incompletely replicated chromosomes are forced to undergo mitosis. A decreasing capacity to complete chromosome replication in egg cytoplasm is the only certain kind of 'stable nuclear differentiation' to have been demonstrated so far by nuclear transplantation. (6) When transplanting nuclei from differentiated or slow-dividing cells, it is beneficial to carry out serial nuclear transfers, so that whole embryos can be prepared from those cells of a partially-cleaved first-transfer embryo that have retained a complete chromosome set. (7) By means of serial transfers, it has been possible to prepare swimming tadpoles with normal cells of several different types from larval intestine and adult frog skin nuclei. This last point proves that genes are not lost, irreversibly changed, or permanently inactivated in the course of cell differentiation; there is no evidence against this conclusion from any nuclear transfer experiments.

It would clearly be of interest to determine the genetic content and developmental capacity of cancer cell nuclei, but this is more difficult to do than it might appear. King and McKinnell (1960) and King and DiBerardino (1965) have transplanted nuclei of cells dissociated from kidney tumours of *Rana pipiens*; McKinnell *et al.* (1969) obtained seven swimming tadpoles from 143 such nuclear transfers. However, there is no proof that the cells from which these nuclei were transplanted were tumour cells and not non-malignant host cells present in the tumour tissue. To be sure that a cancer cell nucleus is transplanted, it would be necessary to clone a single cell, demonstrate that some of the cloned cells are malignant by injecting them into a susceptible animal, and then transplant nuclei from the rest of the cloned cells. Furthermore, it is important to bear in mind that poor nuclear-transplant embryo development, if obtained, might be due to changes undergone during cloning rather than to the malignant state of the cells.

1.6. Summary

The main conclusion to be drawn from the experiments summarized in this chapter is that the nuclei of different kinds of cells in an individual appear to be genetically identical. This conclusion applies primarily to qualitative differences between nuclei (such as a loss or stable change in genes) for which nuclear transplantation provides a very sensitive test. The reservations that need to be made about the generality of this statement are the following. Changes in the numbers of copies of a gene (amplification) have been tested for, but not observed, in only a few cases of protein-coding genes; however, gene amplification does take place in the ribosomal genes of oocytes of some, but not all, species. Changes in the relative positions of genes in a genome have not been tested for with any precision, and might take place in the special case of some antibody-coding genes.

Nuclear transplantation has established that stable qualitative gene changes do not take place in the course, and therefore as a cause, of cell differentiation, but such changes can take place during the terminal stages, and as a result, of cell differentiation, for example, in the case of nuclear extrusion in mammalian erythrocytes. Therefore with the possible exception of antibody-forming cells, there is reason to think that differences between somatic cells are neither caused by, nor dependent on, differences in the informational content of their DNA.

2 *Translational control and message-injection into living cells*

2.1. What is translational control and why might it be important ?

TRANSLATION is the synthesis of a protein on a messenger RNA template and is the third major step in the expression of genetic information. The term 'translational control' is used here, in its broadest sense, to include any situation where the rate at which proteins are synthesized from cytoplasmic messenger RNA is different for two kinds of message in the same cell, for the same message in different cells, or indeed for the same message at different times during the life of a cell. In effect this includes any case where message translation varies in a regulated way and where the existence of a message in a cell does not automatically result in the synthesis of proteins from it at a standard rate. Protein synthesis is a complex process involving several steps, many of which are not well understood. Regulation of the process could be achieved if any one of the interacting components (a) undergoes a change in concentration so as to limit or prevent part of the overall reaction, (b) is secondarily modified, for example by addition of poly A to messenger RNA, or (c) is made to a greater or lesser extent available by compartmentalization within a cell or by association with other molecules.

It is important to distinguish, at this point, between control by message-specific components, and control of an unspecific kind. The unavailability for any of the above reasons, of a message, or message-specific component, would lead to a change in the *kinds* of proteins synthesized, and this could substantially affect the differentiation of a cell. In contrast the unavailability of an unspecific component such as ATP could not in any simple way alter the kinds of proteins made and should have only a quantitative, and therefore developmentally less interesting, effect on the proteins synthesized.

There are two reasons why translational control might be important. First it is possible that gene expression might commonly be regulated at this level. To take an extreme point of view, it is possible to imagine that all cells, having the same genes, would also have each gene transcribed into RNA at the same rate and would contain the same messages. Any differences in the kinds of proteins synthesized by cells would, in this case, result solely from the ability of each cell-type to select which messages to translate. Since the existence of very small numbers of each kind of message in a cell could not easily be detected, this type of proposition (at least in a less extreme form) cannot be immediately eliminated on biochemical grounds. We want to know to what extent the range of proteins synthesized in a cell is determined by control at the level of message translation. The second reason for attaching importance to translational control relates to the possibility of transferring genes from one cell to another. If this eventually becomes possible, it is important to know whether the message for a protein-coding gene can be translated in a foreign cell, or whether it would be necessary to transfer, in addition, some genes for translational factors or other message-specific components.

2.2. Examples of translational control

2.2.1. *Criteria for the recognition of translational control*

Numerous cases have been described where translational control might be involved, but very few where this has been proved. Among the more obvious pitfalls is the confusion of changes in enzyme activity with changes in enzyme synthesis, and the uncritical use of inhibitors. For example, substantial changes in the level of an enzyme activity can be brought about by the provision of cofactors as well as by changes in the composition of an enzyme's immediate environment in the cell; changes in an enzyme's activity cannot be assumed to be in any way connected with changes in its synthesis. If a change in enzyme activity is suppressed by an inhibitor of protein synthesis (such as cycloheximide), this suggests that enzyme synthesis may be the cause of the change in activity; but there is the general objection to experiments of this type, that an inhibitor may well have numerous unknown side-effects in addition to the known effect on account of which it is used. Another source of confusion is that changes in the relative amounts of two enzymes in an enucleate cell may well be due to the differential stability of proteins (review by Schimke and Doyle 1970), and this may not be the same *in vitro* as *in vivo*. Once it is certain that protein synthesis and not enzyme activation is being studied, the involvement of message synthesis must be eliminated. This is usually done by using enucleate cells, or by the application of the drug Actinomycin which suppresses RNA synthesis. In enucleate cells there is the possibility that the messages being studied are synthesized by cytoplasmic genes (e.g. those in mitochondria or chloroplasts). If a change in protein synthesis is observed in cells treated with Actinomycin, it is important to know, though very hard to prove, that the dose given is sufficient to suppress fully the synthesis of each message whose translation is being studied.

Even if all these uncertainties are overcome, there remains the possibility that the unequal translation of two messages might be due solely to differences in the structure of the messages. If the nucleotide sequence of two messages differs in the region responsible for ribosome binding, they would always be translated at different rates. But if the same two messages are translated at different *relative* rates at different times, then some kind of translational control must be involved.

Recognition of these difficulties illustrates the point that even in those organisms most often cited in connection with translational control, it is hard to establish beyond doubt that regulation at this level is really taking place. For example, *Acetabularia*, a large unicellular alga, lends itself to a study of this kind, because its nucleus is easily removed by cutting off the foot end of the stalk, and because the cap which differs in shape from one species to another is regenerated if removed. If a foreign nucleus is grafted to an enucleate stalk, and the cap cut off, then the cap regenerated is of the cytoplasmic and not nuclear species (Hämmerling, 1953). Since cap morphology is determined by nuclear genes, cytoplasmic message is evidently used to make the proteins of the regenerated cap. However, this does not itself prove that the translation of these messages is controlled, since it is possible that all cytoplasmic messages are translated at an equal rate and that the new cap is formed by the rearrangement of proteins and enzyme products within the cell. The strongest case for translational control in *Acetabularia* comes from the work of Zetsche and his colleagues (Grieninger and Zetsche 1972; review by Zetsche *et al.* 1970). During cap regeneration, the rise in activity of the enzyme UDPG-pyrophosphorylase (UDPG-pyr) is related to cap differentiation, but the activity of phosphoglucose–isomerase (PGI) increases in a different manner, which is related to the rate of total protein synthesis. The characteristic increase in activity of each enzyme is dependent on enzyme synthesis since it is inhibited by cycloheximide; it is independent of new gene transcription, since the increase also takes place in enucleate cells, and these enzymes are believed for several reasons to be coded for by nuclear genes. The effect is not due to a differential stability of the enzymes, which has been measured. By elimination, it seems that the increasing activity of cap-related enzymes such as UDPG-pyr. must be due to regulation at the level of translation. The magnitude of this effect is however quite small. The increasing activities of cap-related and other enzymes never get out of step by more than 2–3-fold.

Several circumstances in which translational control is thought to operate in multicellular organisms are outlined below. Eventually it will be possible to classify examples of translational control according to the particular step which is regulated. At present not enough is known about the mechanism or control of translation to do this, but it does seem that delayed translation and message-specific initiation factors are commonly involved.

2.2.2. *Delayed translation*

In several developing or differentiating systems, it has been possible to estimate the times of transcription or translation in relation to the appearance of a new type of enzyme activity or protein synthesis. The period of time over which Actinomycin suppresses new gene expression is taken to show the time of gene transcription, and the cycloheximide-sensitive period to show the time of message translation. As seen in the following example, a gap of several hours is commonly observed between the times of transcription and translation.

The life cycle of the cellular slime-mould, *Dictyostelium discoides* (Acrasiales) involves a series of changes in enzyme activity which have been investigated in detail by Sussman and colleagues (e.g. Sussman 1966; Newell, 1971). For most enzymes (e.g. UDP galactose transferase) the beginning of the Actinomycin-sensitive period precedes the beginning of the cycloheximide-sensitive period by 3–4 hours. In no case has transcription been shown to stop before translation starts. However it appears that this step is regulated because in the same species, it is reduced to 1 hour for the enzyme trehalose phosphate synthetase.

This example does not strictly demonstrate translational control, because the variable gap between transcription and translation may involve the processing or transport, rather than translation, of messages. There are now some indisputable examples of messages existing in cell cytoplasm in an untranslated form. The demonstration of this phenomenon was contingent on the development of a method of translating a putative message into an identifiable protein. Although experiments purporting to achieve this have been reported since 1960, it commonly turned out, on closer examination, that the added RNA was stimulating the translation of mRNA already associated with the ribosomes which were added to the incubate (Drach and Lingrel 1966). It was only in 1969 that Lockard and Lingrel provided the first indisputable evidence for the successful *in vitro* translation of purified mRNA. Mouse haemoglobin mRNA was added to a rabbit reticulocyte lysate (crude cell-free system), and the labelled products were shown to include mouse β-globin which is clearly separable from rabbit α- or β-globin by CM cellulose chromatography. Since then, cell-free systems prepared from several other cell-types, such as chick muscle (Heywood 1969), mammalian Krebs ascites cells (Dobos *et al.* 1971; Mathews and Korner 1971), and mammalian liver (Prichard *et al.* 1971) have been proved to translate added heterologous messages correctly. A substantial purification of the essential components of such systems has now been achieved and a wide range of plant, animal, and virus RNAs has been successfully translated.

2.2.3. *Untranslated cytoplasmic mRNA*

Much the most fully analyzed example of this kind in development concerns fertilization in sea urchins, a subject reviewed by Monroy (1965) and Gross (1967). When an allowance is made for an increase in permeability to precursors, and when contamination of eggs with oocytes or over-ripe eggs is avoided, a 15-fold or greater (according to species) increase in the rate of protein synthesis can be consistently observed when unfertilized eggs are compared to eggs 2 hours after fertilization. Hultin (1961) made the important observation that this same difference in protein synthesis can also be observed with microsomes *in vitro*. That the increase in protein synthesis is independent of new mRNA synthesis was indicated by the experiments of Gross *et al.* (1964) using Actinomycin to suppress RNA synthesis, and by Brachet *et al.* (1963) and Denny and Tyler (1964) using centrifugation to prepare enucleate egg fragments. Actinomycin-treated eggs, and enucleate halves, showed a similar increase in protein synthesis to that seen in normal eggs. Since then numerous differences have been reported between fertilized and unfertilized eggs, affecting, for example, ribosomal characteristics (Maggio *et al.* 1968; Metafora *et al.* 1971), amino acyl-tRNA synthetase activities (Ceccarini *et al.* 1967), and poly-A attachment to RNA (Slater *et al.* 1972). While any one or all of these changes might contribute to the increased rate of protein synthesis, these results are also quite consistent with the view that messenger RNA is present in the unfertilized egg in a masked or unavailable form.

The elimination of several possible explanations for the effect was achieved by Humphreys (1969; 1971), who compared eggs before and after fertilization, and showed that the proportion of total cell ribosomes present in polysomes increases by about 30 times, but that (a) the time for which a ribosome remains on a polysome, and (b) the average size of polysomes (Fig. 11) both stay the same. These findings are inconsistent with any explanation which attributes the lower rate of protein

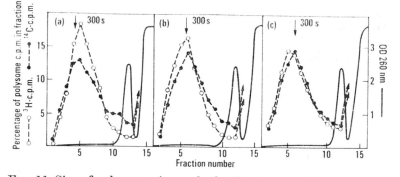

Fig. 11 Size of polysomes in newly-fertilized sea urchin eggs during the activation of protein synthesis. The rate of protein synthesis increases nearly 30 times during the few hours that immediately follow fertilization. Polysomes (which sediment at about 300s) were obtained from eggs which had been labelled with ^{3}H-amino acids for 5 minutes (so as to label nascent protein chains), at the following times after fertilization: (a) immediately, (b) 1 hour, (c) 6 hours. The filled circles (● · · · · · ●) represent ^{14}C-labelled polysomes from more advanced embryos; an equal amount of these polysomes was added to each sample before centrifugation, because the newly formed polysomes do not contribute a significant amount of optical density (——————). (From Humphreys 1971).

synthesis in unfertilized eggs to the lack of a general translational component such as a species of tRNA, an initiation factor, or a ribosome, in fact to the lack of any component which is released from polysomes at the end of each translational event. The provision of any of these components on a scale sufficient to increase protein synthesis by 15 or more times would necessarily result in the formation, at least temporarily, of polysomes carrying a much reduced number of ribosomes. Humphreys was therefore able to say that new messages, on being made available, are almost immediately utilized with complete efficiency. The provision ('unmasking') of mRNA at fertilization must therefore be due either (i) to the removal of an inhibitor associated with message, (ii) to the provision of a component needed to bring this message into use, or (iii) to the release of message from a sequestered site in the unfertilized egg.

Gross *et al.* (1973) have recently provided a direct demonstration of the existence of untranslated messenger RNA in the cytoplasm, but not on the polysomes, of unfertilized eggs. By centrifuging egg homogenates, they obtained some small 30–40s particles (smaller than the 80s monosomes, and much smaller than polysomes) from which RNA was extracted. This RNA was shown to include histone message because it promoted histone synthesis when added to a cell-free protein synthesizing system prepared from ascites cells and because it competed very effectively in the hybridization of labelled histone mRNA to DNA. The labelled histone message used for this test was obtained from a class of small polysomes in fertilized eggs, and was believed for a variety of reasons to consist mainly of authentic histone message (Kedes and Gross 1969). Presumably the histone message contained in the 30–40s particles of un-fertilized eggs is transferred to polysomes soon after fertilization and then contributes, together with newly transcribed histone message, to the elevated rate of histone synthesis stimulated by fertilization. The particles which contain unused messages are of great interest, but have not yet been shown to contain proteins different from those of the mRNP particles contained in polysomes. It is not therefore clear whether these excess messages fail to be incorporated into polysomes because they are associated with a special inhibitory 'masking' protein, or because the eggs have only a limited supply of a component needed to bring messages into translational use. In the former case the messages would be truly 'masked'; in the latter case they would merely be present in excess of the cells' trans-lational capacity. If sea urchin eggs possess a spare trans-lational capacity of the kind known to exist in frogs' eggs (p. 61), this would argue rather strongly for the existence of messages in a masked rather than excess-of-capacity condition.

A question of special relevance to development is whether the new messages brought into use at fertilization are of the same kind (code for the same proteins) as those in use in un-fertilized eggs. This concerns the general question of whether translational control can be selective with respect to types of messages, and therefore lead to differences in the *kinds* of proteins synthesized in a cell. General analyses of all proteins synthesized before and just after fertilization have not revealed clear differences (Gross 1967). Two identified proteins which are synthesized in significant amounts after fertilization are histones (Kedes *et al.* 1969) and microtubule protein (Raff *et al.* 1972). When Actinomycin is used to inhibit RNA synthesis in fertilized eggs, it has no effect on microtubule protein synthesis, but causes some reduction in the normal rate of histone synthesis, presumably because new message for histone but not for microtubular protein is normally synthesized. However, in neither this nor any other aspect of sea urchin fertilization is there any firm reason to suppose that maternal messages for different proteins are handled in different ways after fertilization. Sea urchin fertilization has therefore provided a clear example of translational control with a pronounced effect on the *amounts* of proteins synthesized, but not necessarily with any effect on the *kinds* of proteins made.

The germination of plant seeds constitutes another possible example of masked messages. An ungerminated seed, which contains a multicellular embryo and cotyledon, ceases to synthesize protein at a detectable rate, when it is released from the parent plant and becomes desiccated. As soon as the dried seed is allowed to inbibe water, protein synthesis increases rapidly, new polysomes are formed, and the activity of numerous enzymes greatly increases (Marré 1967). It appears that this activation of protein synthesis involves the binding of mRNA, already present in ungerminated seeds, to ribosomes to form polysomes. Using wheat seeds, Marcus and Feeley (1966) showed that ribosomes from ungerminated seeds could be activated *in vitro* by incubation at 30° after addition of ATP and Mg^{2+}, to a level of protein synthesis commensurate with that of germinated seeds (Table 5). Chen *et al.* (1968) have presented evidence for the existence of conserved mRNA in ungerminated seeds, and have shown by competitive hybridization, that the ungerminated seed already contains those kinds of RNA which are synthesized at a very low level after germination. Hence it is likely that the initial rise in protein synthesis following inhibition involves translational control. However, as in sea urchin fertilization, there is no reason to suppose that new *kinds* of proteins are synthesized as a result, and therefore that any discrimination between messages takes place.

The existence of unused or 'masked' cytoplasmic messages will probably turn out to be quite common in cells. Globin mRNA-containing particles have been found in reticulocytes and there are several other conditions in which masked messages may exist (see reviews by Spirin 1966, and Tyler 1967).

TABLE 5

In vitro *activation of ribosomes from ungerminated wheat embryos*†

Experiment	Conditions of preincubation (before isolation of ribosomes and polysomes)	Incorporation into protein in cpm/mg (a measure of mRNA bound to ribosomes)
1	None	140 (3% of max.)
2	12 min, 30°C, +Mg²⁺, +ATP	5140 (100%)
3	″ ″ +ATP	2190 (43%)
4	″ ″ +Mg²⁺	285 (5%)
5	″ 0°C, +Mg²⁺, +ATP	490 (9%)

† (From Marcus and Feeley, 1966.) Ungerminated wheat seed embryos were homogenized, and a low-speed pellet collected. In Experiment 1, ribosomes and polysomes were isolated from the low speed supernatant and tested for amino acid incorporating activity in a cell-free system. In Experiments 2–5, the low-speed supernatant was 'pre-incubated' under the conditions specified; ribosomes and polysomes were then extracted and assayed for incorporating activity as in Experiment 1. The amount of incorporation reflects the amount of polysomes (i.e. mRNA bound to ribosomes). The results imply that mRNA is not associated with the ribosomes present in the homogenate (of ungerminated seeds), and that it becomes attached to ribosomes during preincubation. This *in vitro* process simulates the events which normally take place at germination *in vivo*.

2.2.4. *Control at the initiation of translation*

There is no doubt that the types of messages whose translation is currently best understood are those for α and β globin contained in the haemoglobin of mammalian reticulocytes. Approximately ninety per cent of the proteins synthesized by adult rabbit reticulocytes are α and β-globin chains; these proteins are synthesized in almost equal amounts, the amino acid sequence of each is known, and factors affecting globin synthesis have been extensively investigated (reviews by Baglioni 1963, and London *et al.* 1967). Special interest in the handling of the messages for these two proteins stemmed from the finding of Hunt *et al.* (1968) that β-globin is synthesized on larger polysomes (average 4–5 ribosomes) than α-globin which is made on polysomes containing an average of 3–4 ribosomes (Fig. 12). Since as many α chains are synthesized as β chains per time, the smaller number of nascent α chains present on polysomes must be compensated for either by a faster passage of ribosomes along α messages, or by a greater abundance of α messages each of which receives an attached ribosome less frequently. It now seems clear that the second explanation is correct. This was shown in two ways. Inhibitors of protein synthesis which greatly slow down the rate of elongation, and therefore also of initiation, cause the ratio of α: β chains synthesized to be determined by the relative abundance of α and β messages in the cell; under these conditions the ratio of α: β globin synthesis was raised to about 1·4:1 (Lodish 1971). Lodish and Jacobsen (1972) subsequently demonstrated that the elongation and termination rates for α and β globin synthesis are the same. Evidently normal reticulocytes contain substantially more α than β globin mRNA, but β mRNA is initiated for protein synthesis more frequently than α mRNA (Fig. 12). The reason for this curious combination of regulation at the translational and transcriptional levels, the end result of which appears to be the same as if neither step were regulated, is not understood. It could be connected with the fact that α and β chains are not always synthesized coordinately; during foetal life α but not β chains are synthesized (review by Marks and Kovach, 1966).

This example is important for the present discussion because it demonstrates clearly that the translation of two kinds of messages can be initiated at different rates in the same cell. It may well be that this results from a difference in the coded initiation sequences of the two messages, and it does not necessarily imply the existence of message-specific initiation components. There is however some direct evidence for such components in experiments with cell-free systems in which messages from one cell-type are translated with components of another cell-type.

The first well-documented evidence for message-specific translational components has come from Heywood (1969), who has added chick myosin message to cell-free systems composed of ribosomes, etc. from reticulocytes or muscle cells. It was found that chick myosin mRNA was translated only poorly if at all in a chick erythroblast (or rabbit reticulocyte) cell-free

48   *Translational control and message-injection into living cells*

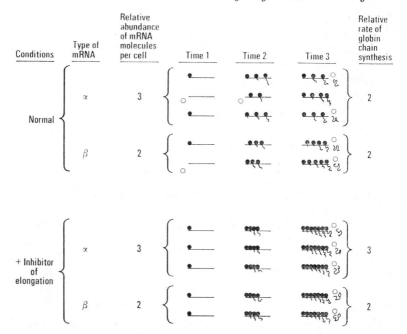

FIG. 12 Translation of mRNAs for α and β globin in rabbit reticulocytes. Each solid horizontal line represents a molecule of α or β mRNA. Free ribosomes are shown by open circles, and ribosomes in a polysome by closed circles. The twisted lines attached to mRNA molecules show nascent proteins still present in a polysome, or complete globin chains just released from a polysome. In the upper figure (normal development), a ribosome takes about 35 seconds to move along the whole of an mRNA molecule and to synthesize one globin chain; times 1, 2, and 3 therefore represent the state of a polysome at about 5, 25, and 40 seconds after the first ribosome has attached to mRNA. In the lower figure translation has been slowed down to about one-third of its normal rate by an inhibitor of protein synthesis. In this case a ribosome would take about 1½ minutes to move along an mRNA molecule, and the times 1, 2, and 3 represent the state of polysomes at about 5, 60, and 100 seconds after the start of translation. For simplicity, each diagram shows the attachment of ribosomes as a message starts to be used for the first time. (Based on Hunt *et al.* 1968; Lodish 1971, and Lodish and Jacobsen 1972).

system, unless accompanied by initiation factors from muscle cell ribosomes (Table 6); the equivalent factors from erythroblast or reticulocyte ribosomes had no effect on myosin mRNA

TABLE 6

Effects of ribosome factors on the in vitro *translation of mRNA with ribosomes from different cell-types†*

Assay system and ribosome source	Ribosome factors	Myosin mRNA	Incorporation
Chick muscle ribosomes (complete)	—	+	1400
" " " (complete)	—	–	155
" " " (KCl washed)	—	+	145
" " " (")	Muscle	+	1250
" " " (")	Erythroblast	+	185
Chick erythroblast ribosomes (complete)	—	+	20
" " " (KCl washed)	—	+	0
" " " (")	Muscle	+	1040
" " " (")	Erythroblast		25
Rabbit reticulocyte lysate	—	+	5
" " "	Muscle	+	610
" " "	Erythroblast	+	10

† Chick myosin mRNA is added to a cell-free system which contains ribosomes (and other components) from various cell-types, and which is supplemented with factors released from ribosomes by 1·0 M-KCl. Myosin was identified by gel electrophoresis. (From Rourke and Heywood 1972.)

translation. Recently Rourke and Heywood (1972) have greatly strengthened the earlier results by using more stringent criteria (including peptide analysis) to identify synthesized myosin and haemoglobin. Very few other experiments have so far been reported, where as in those of Heywood, factors from two cell-types have been tested in cross combinations. The substantial stimulation of message translation often observed when homologous initiation factors are supplied (e.g. Metafora *et al.* 1972) may mean only that the cell-free system used was deficient in these factors and these are not therefore shown by such experiments to be message-specific. However, Heywood's conclusions have received support from other work implicating message-specific initiation factors (e.g. Cohen 1971; Prichard *et al.* 1971; Ilan and Ilan 1971), and especially from the purification by Wigle and Smith (1973) of an initiation factor which is required for EMC virus RNA to be translated in an ascites cell-free system, but which has no effect on the translation of globin mRNA in the same system.

The specificity of translation revealed by Heywood's experiments is not complete in so far as messages from one cell-type can nearly always be translated *to some extent* by ribosomes etc. from another cell-type. It could be argued that the specificity of translation may have been reduced or lost by the purification of ribosomes and other components of a cell-free system which will not necessarily reveal *in vitro* the kinds of mRNA which would be translated in living cells. Nevertheless, even systems consisting of unpurified components, such as whole cell lysates, can translate messages from quite different cell-types. Reticulocyte lysates are most revealing since they are derived from a specialized and pure cell-type, and these have been able to translate mRNA for immunglobulin light chains from mouse myeloma cells (Stavnezer and Huang 1971), for ovalbumin from hen oviduct cells (Rhoads *et al.* 1971), and for α-crystallin from calf lens (Berns *et al.* 1972*a*). More detailed information on the translation of mRNAs in cell-free systems can be obtained from a review by Lingrel (1974). The conclusion that can be drawn from the translation of messages in heterologous cell-free systems is that specialized cells possess message-specific components (probably initiation factors), but also unspecific factors enabling them to translate messages from other cell-types. These results suggest that message-specific initiation factors may well have the effect of conferring a translational advantage on messages characteristic of specialized cell functions.

2.2.5. *The importance of translational control as judged from the experiments outlined*

In none of the examples cited so far has it been shown that the translation of two kinds of messages in the same cell can be regulated independently. We know only that initiation factors can distinguish the messages of different cell-types, such as those for myosin and haemoglobin, which probably do not normally co-exist in the same cell. The differential usage of two messages in the same cell is suggested by experiments of Gelehrter and Tomkins (1969) using an established line of rat hepatoma (liver cancer) cells which synthesize tyrosine aminotransferase (TAT). Labelled antibodies were used to measure TAT synthesis. If cells are grown in the presence of hormone (which induces enzyme synthesis partly at the transcriptional level), and their medium then supplemented with serum, the rate of TAT synthesis increases by more than two fold and most of this increase is unaffected by Actinomycin, but is inhibited by cycloheximide. If attention is paid to the Actinomycin-insensitive stimulation of enzyme synthesis, that is the increase due to translation, then serum increases TAT synthesis by over 90 per cent, but the synthesis of other proteins by only 23 per cent. Evidently serum has a differential effect on the translation of different messages in the same cell (Table 7). The role of translation and transcription in regulating TAT synthesis in this cell line has been reviewed by Tomkins *et al.* (1968).

TABLE 7

Control of enzyme synthesis at the level of translation
(From Gelehrter and Tomkins 1969.)[†]

	Synthesis (c.p.m./mg protein)		
	Without serum in medium	With serum in medium	Stimulation due to serum
Tyrosine aminotransferase synthesis[‡]	600	1180	197%
Total cell protein synthesis	295 000	362 000	123%

† Cultured rat hepatoma cells, which have been induced by hormone to synthesize tyrosine aminotransferase, are grown with or without serum in the medium.

‡ Assayed by precipitation with radioactively-labelled antibodies prepared against purified tyrosine aminotransferase enzyme.

When trying to assess the importance of translational control, we should note two general characteristics of the examples cited above. Translational control has so far been shown to operate in development only in such a way as to affect the rate of translation of *all* messages in a cell. It has not yet been shown that animal egg fertilization or plant seed germination involves the usage of new *kinds* of messages which were not already making protein. In differentiated cells we have seen that the translation of α and β globin message is initiated at different rates, but it is not known whether the *relative* rate of α and β message translation can be regulated. In cultured hepatoma cells two kinds of messages in the same cell can be translated at different *relative* rates according to conditions, and this shows that it is possible for translational control to have qualitative effects on the kinds of proteins synthesized. Nevertheless the magnitude of such effects is small. Cells already induced to synthesize TAT can be made to do so at a relatively faster or slower rate by a serum-activated translational mechanism, and initiation affinity accounts for a 30 per cent difference in the numbers of α and β chains synthesized per cell.

Although much valuable information has emerged from the description of translation in normal cells and from the use of cell-free systems for translating added messages, these methods have certain limitations affecting the kinds of conclusions which they can eventually yield. We now discuss in some detail another type of experimental approach which, it is hoped, may combine the advantages of working on living cells with the experimental advantage of working on pure messages from different cell-types.

2.3. The experimental analysis of translation in living cells

The experimental systems to be discussed involve the transfer of mRNA from one cell to another, and there are several reasons why it is useful to attempt to do this. To carry out a detailed investigation of translational control, it is desirable (1) to work with the messenger RNAs for *known* proteins, (2) to eliminate synthesis of the messages whose translation is being studied, and (3) to be able to vary independently the concentration of each translational component. Requirements (2) and (3) cannot be met conveniently, if at all, in unmanipulated normal cells. Cell-free systems can in principle satisfy these demands, but they may fail to provide a meaningful test of normal translational control processes for the following reasons. First, it is possible that the specificity of cell components may be altered when they are extracted and purified, a point already emphasized. Second, components extracted

from a cell and mixed together in a cell-free system may not
have been free to interact in the living cell, since the distribution
of some cell components is certainly compartmentalized.
Lastly it is impossible to know at exactly what concentration
each component should be supplied in a cell-free system to
simulate *in vivo* conditions. This is a problem which affects not
only ribosomes, initiation factors, etc., but also the inorganic
ingredients of the medium. For example, a doubling of the
Mg^{2+} concentration in a reticulocyte cell-free system can alter
by a factor of two the ratio of $\alpha:\beta$ globin synthesized from
added mRNA. Even if the overall ionic composition of a cell is
known, it is possible that local differences exist within the cell,
and in this case it would be impossible to be sure of reproducing
in vitro the normal ionic conditions.

For all these reasons there is a very strong incentive to
transfer messages from one cell to another. If this could be
done successfully, it would eliminate all uncertainties about the
ionic composition of a cell-free system, and about changes
in the specificity of extracted cell components. Messages
successfully transferred to a living cell will be translated under
conditions, and by components, which are necessarily normal
for the cell being tested. Such a system would also satisfy the
three major requirements specified above.

2.3.1. *The translation of mRNA entering cells by diffusion or injection*

Some of the first experiments which demonstrated an effect
of exogenous RNA on animal cells were those of Niu and
collaborators (e.g. Niu *et al.* 1961). In some cases (e.g. Niu and
Deshpande 1973), a property (such as pulsating myoblasts)
characteristic of the donor tissue from which RNA is extracted
(chicken heart) is induced in the recipient tissue (post-nodal
chick blastoderm) to which RNA from heart but not other
tissues is added. However there has been some general reluc-
tance to fully accept the results of such experiments, partly
because the specificity and reproducibility of the results has
been questioned (e.g. Hillman and Hillman 1967), and partly
because the experiments did not include the characterization
of a protein synthesized by the donor and RNA-treated cells,
but not synthesized by untreated cells. Some of these objections

have been overcome in more recent work (e.g. Grässman and Grässman 1971; Maeyer-Guignard *et al.* 1972; Tuohimaa *et al.* 1972), and in a series of experiments involving uptake of putative antibody mRNA (referred to in Wang *et al.* 1973); the subject has been reviewed by Bhargava and Shanmugam (1971), Lane and Knowland (1974), and Niu (1974). Nevertheless, in all these cases, the amount of RNA which enters a cell is insufficient to permit a purification and full biochemical characterization of the protein supposed to be coded for by it. It therefore has yet to be formally proved that the stimulated protein synthesis or enzyme activity results from the translation of exogenous message rather than from the activation of host cell genes or enzymes by a component of the message preparation. This uncertainty has been overcome by injecting mRNA directly into cells.

The successful injection of messenger RNA into cells was first carried out on oocytes of *Xenopus* by Gurdon *et al.* (1971) and Lane *et al.* (1971), using a procedure originally developed to inject nuclei into oocytes (Gurdon 1968*b*). The method involves the construction of micropipettes calibrated to deliver a known volume in the range of 10–100 nl (see Appendix C). An ovary is removed from a frog, and divided into small groups of about 5 large oocytes each of which is injected with mRNA (Fig. 13). Each group is usually used 'unplucked' (i.e. without the removal of the injected oocytes and their surrounding follicle cells from the small uninjected oocytes and ovarian tissue); 'plucked' oocytes permit a more accurate assessment of the amount and kinds of proteins synthesized from injected messages. The cluster is then cultured in a small volume of incubation medium (Appendix C) containing a labelled amino acid. Injected oocytes can be cultured for many days, and labelled for any desired period during this time. Amino acids are incorporated very efficiently, and if one with a small intracellular pool is used, such as methionine, over 10^6 cpm can be incorporated into protein by each mature oocyte within a few hours. Unfertilized eggs can be used for these experiments, but have to be labelled by microinjection rather than by incubation (Gurdon *et al.*, 1971).

Proof that the injected mRNA is correctly translated comes principally from the work of Lane using messages for rabbit,

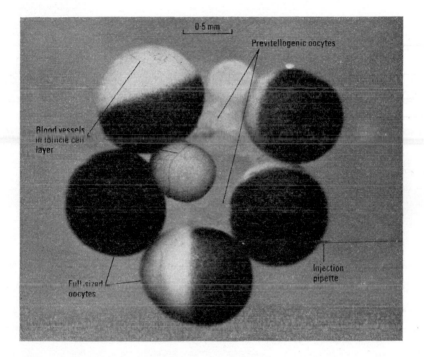

Fig. 13 A cluster of oocytes of *Xenopus laevis*, with injection pipette. The follicle cells which surround each oocyte (*cf.* Fig. 3) are too small to be seen, though small blood vessels in the sheet of follicle cells are visible. The injection pipette has an external diameter of about 20μm.

mouse, and duck haemoglobins. Oocytes were injected with reticulocyte 9s RNA of each species, they were labelled with ^{3}H-histidine overnight, and the proteins contained in the homogenized samples analyzed by the procedures listed in Table 8. Clearly oocytes synthesize globin which is indistinguishable by many criteria from the kind of globin coded by the injected mRNA. The homogenization medium contains unlabelled haemoglobin, subunits of which appear to exchange with the globin chains synthesized in oocytes. It is not essential to add haemin to the injected mRNA; the effects of haemin are described on p. 60. Control oocytes injected with saline injection medium but not with Hb mRNA, or injected with any kind of RNA other than reticulocyte 9s RNA, synthesize no detectable amounts of haemoglobin (Lane *et al.* 1971). The fact that oocytes always synthesize the kind of rabbit, mouse, or duck globin characteristic of the species donating the mRNA (Table 8) constitutes compelling evidence that the injected messenger RNA is translated. *Xenopus* tadpoles synthesize two kinds of haemoglobin, which are different from the two kinds of haemoglobin synthesized by adult *Xenopus* (Maclean and Jurd 1971), but neither has been characterized in detail. If tadpole genes were activated by the injected RNA, then tadpole globin should be synthesized after injection of each kind of mammalian mRNA. Evidently the injected mRNAs are translated into the proteins they code for; the only reservation is that while the proteins synthesized in response to injected mRNA have been analyzed in considerable detail (Table 8), it is conceivable that small translational errors would not have been seen. *Xenopus* oocytes can therefore be said to have provided the first conclusive evidence that mRNA from one kind of cell can be translated in the living cytoplasm of a cell of a different type and species. This is an important conclusion since it argues decisively against the view that a cell can translate only those messages which it would normally translate and to which its ribosomes or other translational components might have been preadapted.

2.3.2. *The specificity and efficiency of translation in oocytes*

If message injection into oocytes is to be used for a detailed analysis of translational control, we must know whether mRNAs

TABLE 8

Evidence for accurate translation of globin mRNAs injected into frog oocytes[†]

Species of Hb mRNA injected	Fractionation method	Result of fractionation (Oocytes labelled with ³H-his; carrier Hb labelled with ¹⁴C-his, or recognized by optical density (OD)).	Similarities between oocyte-derived and carrier Hb
Rabbit, duck, and mouse	Sephadex G-100 in tris-glycine buffer	Retarded ³H-peak coincident with OD of carrier Hb	Overall size of Hb
Rabbit	Polyacrylamide gel electrophoresis in tris-glycine buffer	Sharp ³H-peak coincident with OD of carrier Hb	Overall size and charge of Hb
Rabbit, mouse and duck	CM cellulose chromatography of free globin chains; (elution conditions will separate globins of rabbit, mouse, duck and frog.)	Coincidence between all globins of donor species and oocyte-derived materials.	Overall charge of each major globin species
Rabbit	Gradient elution of tryptic peptides from cation exchange resin	Coincident elution, and similar specific activity, of ³H-oocyte-derived and ¹⁴C-carrier peptides (for 7α and 8β globin peptides)	General similarity in charge for about half of all tryptic peptides in each kind of globin
Rabbit and duck	Paper electrophoresis at pH 3·5 and pH 6·5, and paper chromatography.	As above for 7 peptides from both α and β rabbit globins, and for 5 or 6 peptides of each major duck globin.	

† Details of these results may be found in Lane *et al.* (1971), Marbaix and Lane (1972), Lane *et al.* (1973), and Gurdon *et al.* (1974).

from all different species and cell-types can be translated in frog oocytes, and whether injected messages are translated at an efficiency comparable with that observed in normal cells. The different kinds of messages which have been proved so far to be translated in frog oocytes are listed in Table 9, and the following conclusions can be drawn.

Regarding species specificity, there is no obvious restriction of translation between different vertebrate groups, and the results with bee promellitin mRNA suggest that some invertebrate messages may also be translatable. So far, no bacterial or bacteriophage mRNA has been translated in frog oocytes in spite of persistent attempts with RNA phages such as f2, R17, and Qβ (Gurdon *et al.* 1971; Lane and Knowland 1974). Among the animal viruses, a picorna virus (EMC) virion RNA is translated, but attempts to translate virion RNA from RNA tumour viruses such as AMV and Rauscher murine leukaemia virus have not so far been successful. Rollins and Flickinger (1972) have found a stimulation of collagen synthesis in oocytes injected with tadpole RNA. If this effect is due to the translation of injected messages, rather than to the increased translation of oocyte messages, it would imply that oocytes can translate messages from amphibia as well as from other vertebrates. Concerning cell-type specificity, the results in Table 9 show that mRNAs from several different specialized cells, in addition to reticulocytes, can be successfully translated in oocytes. We conclude that the translation of injected messages in oocytes is not restricted by any absolute cell-type specificity, and probably by no species-specificity, among vertebrate animals. We discuss now the separate and important possibility that messages from different cell-types and species, while all translated, are not all translated with the same efficiency.

The efficiency of translation can be most accurately calculated for rabbit or mouse Hb mRNA, when used at a sub-saturating concentration (see p. 61). Earlier calculations (Gurdon *et al.* 1971), which were based on the number of 9s RNA molecules injected and the number of Hb molecules synthesized, are consistent with the results of more recent experiments which show that each molecule of injected 9s RNA yields an average 15–20 globin molecules per minute.

TABLE 9

Different kinds of mRNA successfully translated in injected
Xenopus *oocytes*

Species and type of mRNA and of product synthesized in oocyte	Fractionation and identification	Reference
Rabbit, mouse, and duck haemoglobin (α and β globins)	see Table 8	Table 8
Calf lens crystallin (αA2 crystallin)	SDS and acid urea gel electrophoresis of crystallin complex; basic urea gel electrophoresis; paper chromatography of tryptic peptides	Berns *et al.* (1972*b*)
Mouse myeloma light chain	Antibody precipitation; sephadex separation, and gel electrophoresis of light chains. Analysis of tryptic peptides.	Stevens and Williamson (1972). Smith *et al.* (1974).
Encephalomyocarditis virus of mouse (stable proteins)	SDS gel electrophoresis of 6 stable viral proteins; electrophoresis and chromatography of major peptides of 3 stable viral proteins.	Laskey *et al.* (1972)
Trout testis protamine	Biogel fractionation of sulphuric acid extract; acrylamide and starch gel electrophoresis, chromatography on CM cellulose; electrophoresis of tryptic peptides.	Wu, Dixon and Gurdon (unpublished)
Honeybee promellitin	Paper chromatography of butanol/ammonia extract; electrophoresis of peptic digest; analysis of tryptic digest of peptic fragments.	Lane and Knowland (1974)
Collagen of 3T6 cultured mouse cells	Quantitation of labelled hydroxyproline and proline in acid-insoluble extract. (Assay carried out on unfertilized eggs, because oocytes but not eggs have background of hypro-labelled protein).	Lane and Knowland (1974); Knowland *et al.*, (1974).
Frog tadpoles (*Rana pipiens*)	As above, but assay on oocytes.	Rollins and Flickinger (1972)

† In each case proteins were labelled with a single amino acid present in about half of all tryptic peptides; synthesized proteins were identified by following their radioactivity.

We now know that this figure needs to be corrected in two respects. β Globin mRNA is translated about 5 times more efficiently than α globin mRNA (below); 35 per cent of the mRNA sample used was not functional because it lacked the usual poly-A component, and was not translated in a cell-free system (Morrison *et al.* 1972); only 65 per cent of the 9s RNA injected is functional message. After making the appropriate corrections we conclude that each molecule of functional β mRNA yields about 30 molecules of β globin per minute; this compares to the production of about 110 molecules of β globin per mRNA molecule per minute in rabbit reticulocytes incubated at 19°C (Hunt *et al.* 1969). However it is by no means certain that every functional mRNA injected into an oocyte is used for translation, and if not, it is possible that β mRNA is translated as efficiently in injected oocytes as in the normal reticulocytes from which it was taken. In conclusion, it is clear that β globin RNA injected into living oocytes is translated at an efficiency which is within the range of that observed in cells where the message normally functions. The translation of injected mRNA also appears to proceed by the normal mechanism, since it is inhibited by puromycin (Gurdon *et al.* 1971), and is carried out on polysomes (Lingrel and Woodland 1974).

As emphasized at the beginning of this chapter, a question of obvious developmental importance is whether the same two messages can be translated with different relative efficiencies in two cell-types. This phenomenon is illustrated by comparing the translation of α and β globin mRNA in mammalian reticulocytes and frog oocytes (Gurdon *et al.* 1974). If Hb mRNA is extracted from rabbit (or mouse) polysomes and translated in a cell-free system prepared from the same cells, the ratio of α and β globin synthesis is about 8:10. However when the same preparation of rabbit (or mouse) Hb mRNA is injected into oocytes, the ratio of α:β globin synthesis is only 2:10 (Fig. 14). This effect could have been due to the existence of different saturation levels for α and β mRNAs, but this is not so because the same ratio of α to β globin synthesis is observed when the mRNA is injected above and below saturation (Gurdon *et al.* 1974). The different behaviour of globin mRNAs in oocytes and reticulocytes illustrates a type of translational control not already demonstrated by the different efficiency with which α

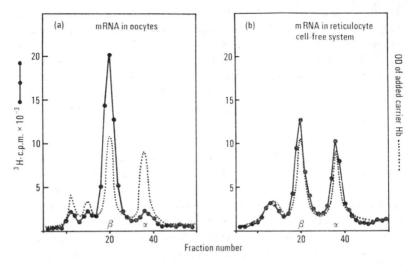

FIG. 14 The relative efficiency of translation of mouse α and β globin mRNAs in frog oocytes and in a rabbit reticulocyte cell-free system. The same sample of mRNA was used in both experiments. In a reticulocyte cell-free system (Fig. b), α and β mRNAs are translated with about equal efficiency, but in living frog oocytes, α mRNA is translated only about 20 per cent as efficiently as β mRNA. Carrier α and β globin (· · · · · ·) was added to each sample before extraction and shows that there was no differential loss of α and β globin during these analyses (From Gurdon *et al.* 1974).

and β mRNAs are translated in reticulocytes. The latter effect could be entirely due to differences in the coded initiation sequence of these messages, and does not necessarily demonstrate translational control. The fact that the same messages are handled differently in oocytes and reticulocytes shows very clearly that the use of two messages in the same cell can be controlled by a mechanism operating at the translational level.

The basis of this effect is partly understood. Giglioni *et al.* (1973) have shown that the translation of α-globin mRNA in oocytes is dramatically increased by haemin. When mRNA is injected at a subsaturating concentration, the simultaneous injection of haemin (to give a final intracellular concentration of 30 μM) results in a nearly equal rate of α and β globin synthesis. The following findings from the writer's laboratory throw some light on the mode of action of haemin. It is effective if supplied two days after an initial injection of globin mRNA alone. It does not work solely by increasing the size of α mRNA polysomes. A non-haemin component is formed in *Xenopus* embryos which copies the haemin effect for rabbit, but not for mouse, α mRNA. In conclusion, haemin is clearly an example of a factor capable of regulating translation in a message-specific way.

2.3.3. *Message saturation experiments and spare translational capacity*

A message saturation experiment relates the amount of mRNA injected into an oocyte to the amount of its translational product synthesized. It is important to obtain this information for three reasons.

First, a saturation experiment can show whether injected messages are translated only at the expense of endogenous messages, or whether a cell has spare translational capacity. Secondly, such an experiment defines the amount of mRNA needed to saturate an egg or oocyte, and therefore provides essential background information whenever it is necessary to know (as it nearly always is) whether injected messages are being translated under conditions in which mRNA is, or is not, the rate-limiting component. Thirdly, the extent to which different messages compete against each other can be investigated directly only if a cell can be saturated with at least one of the messages being tested.

The result of a typical saturation experiment with oocytes is shown in Fig. 15a. Regarding spare translational capacity, the results first reported by Moar *et al.* (1971) have now been confirmed in several experiments. At low doses of injected mRNA, the amount of Hb synthesized increases linearly with the amount of message injected, and the total protein synthesized by cells also increases. Most significant is the fact that endogenous messages continue to be translated at the same rate. These experiments have established for the first time in a living cell the existence of spare translational capacity, which is possibly provided in eggs to handle new messages made available by unmasking (p. 44) in early development. The results also prove that the rate of protein synthesis is limited in these cells by the supply of mRNA.

The amount of mRNA needed to saturate an egg or oocyte is

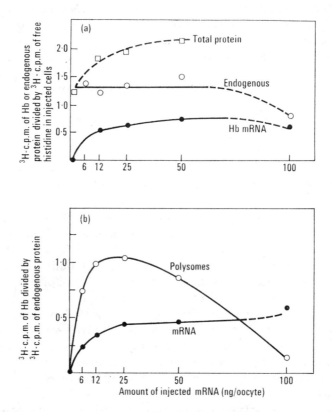

FIG. 15 Saturation experiments with Hb mRNA and polysomes in oocytes. Oocytes are injected with increasing amounts of mRNA or polysomes, and then incubated in ³H-histidine for 12–18 hours at 19 °C. Fig. 15(a) gives an estimate of the actual rate of protein synthesis by relating protein synthesis (as judged by ³H-histidine incorporation) to the amount of unincorporated ³H-histidine in the cells. Fig. 15b shows the ratio of Hb synthesis to endogenous synthesis. Since Hb contains twice as many histidine residues as do average animal proteins of similar size, the actual ratio of protein synthesis (as opposed to histidine incorporation) is about half that shown. These results show that oocytes are saturated by about 10–15 ng of mRNA, and that, at the saturation level, total protein synthesis is increased by nearly 50 per cent. Fig. 15b shows that cells saturated with polysomes synthesize 2–3 times more Hb than similar cells saturated with mRNA. At high concentrations the injected polysomes are toxic. (From Gurdon and Marbaix, unpublished results.)

about 10 ng (or about 3×10^{10} molecules). When saturated with message, the amount of Hb synthesized by injected cells is 30–50 per cent of their total endogenous protein synthesis (i.e. total protein synthesis is raised by 30–50 per cent). It is also of interest that, in eggs and oocytes, less than 5 per cent of all ribosomes are present on polysomes. As development proceeds, free ribosomes are progressively mobilized, so that in stage 40 tadpoles about 75 per cent of all ribosomes are engaged in protein synthesis (Woodland unpublished results).

The third result to emerge from a saturation experiment concerns the ability of different messages to compete against each other. As pointed out by Moar *et al.* (1971), an oocyte which is saturated with Hb mRNA shows very little increase in Hb synthesis, or decrease in endogenous protein synthesis, even when several times the saturating amount of message is injected (Fig. 15a). Subsequent experiments have confirmed this effect, and failed to reveal any substantial competition between the excess Hb mRNA and endogenous messages. Only at very high concentrations of injected mRNA is endogenous synthesis somewhat reduced, probably because of a generally toxic effect of injecting large amounts of RNA.

The lack of competition between messages can be accounted for in two fundamentally different ways. The existence of message-specific factors needed for messages to be brought into use could readily account for the lack of competition between two kinds of messages. However the lack of competition could equally well be accounted for if each mRNA molecule has to be 'activated' by a factor which need not be message-specific so long as it is also not exchangeable between messages. For example, if such a factor were to convert a messenger into a circle, this would ensure that exchangeable components, such as ribosomes (or subunits), would be enormously more likely to associate with the beginning of the same message, than with the beginning of any message which would necessarily be further away. These proposals are illustrated diagramatically in Fig. 16. If we are to understand the handling of messages in development, it is important to know which of these types of mechanism operates, and the following experiments have been carried out to investigate the phenomenon further.

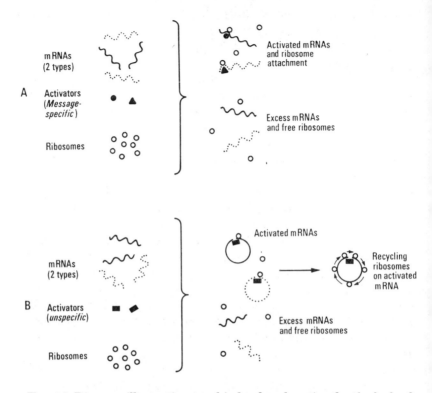

FIG. 16 Diagram illustrating two kinds of explanation for the lack of competition between mRNAs in a living cell. In Fig. 16A, the lack of competition is accounted for by assuming that message-specific initiation factors exist. Therefore two kinds of mRNAs whose translation is dependent on different factors can not compete against each other. In Fig. 16B, the lack of competition between different messages in a cell is accounted for by assuming that an mRNA molecule cannot be translated unless first complexed with a cell component which binds permanently to it. This component would not be released with ribosomes, and the 'activated' message would be stabilized if it were converted into a circle, so that as soon as one round of translation is completed, a released ribosome could be at once reutilized on the same message.

2.3.4. *The lack of competition between messages in the same cell*

First, an attempt has been made to identify the cell component which limits the translation of injected mRNA at saturation, by supplementing a saturating dose of injected mRNA with other cell components. It is found that neither total transfer (tRNA) of reticulocytes or of cultured cells, nor met-tRNA^{met} or met-tRNA^{fmet} of mammalian cultured cells, raises the rate of Hb synthesis in oocytes above the usual saturation level (unpublished experiments of Laskey and Gurdon). The material released from ribosomes by 0·5 M KCl contains initiation factors, some of which may be message-specific (see p. 49). KCl factors from oocyte and reticulocyte ribosomes have been tested with Hb mRNA below saturation, when they might increase the efficiency of translation of messages already being used, and above saturation when they could bring into use those mRNA molecules which are in excess of the cells' capacity. KCl factors had only a small beneficial effect on Hb mRNA above saturation, and no effect below saturation (Marbaix and Gurdon 1972). It is therefore unlikely that either of the two strongest candidates for message-specific translational components, tRNA and KCl factors, is responsible for limiting the translation of Hb mRNA at satur ation, nor presumably for the lack of competition between messages. Some worthwhile information about the nature of the limiting component has come from saturation experiments with polysomes. The result, shown in Fig. 15 b, is that oocytes saturated with polysomes synthesize up to three times more Hb than do cells of the same batch saturated with mRNA. This effect seems not to be connected with the finding referred to above that message-injected oocytes synthesize only 20 per cent as much α as β globin, because an increase in α globin synthesis, so that it would be equal to β synthesis, would only raise the level of Hb synthesis by less than two times. In contrast to experience with mRNA, unpurified polysomes appear to be strongly toxic at high concentrations, and the component which limits their translation at saturation seems to be tRNA. We can conclude that the component which limits mRNA translation is probably present in polysomes. The results with tRNA and initiation factors are consistent with Fig. 16B.

Another approach to the analysis of translational control in oocytes involves the examination of polysomes engaged in translating injected messages. By analogy with the experiments of Lodish (1971) and Humphreys (1971) outlined above (pp. 47, 42), the injection of saturating amounts of Hb mRNA should have one of two effects on polysome size. If the translation of Hb mRNA at saturation is limited by the supply of a component which undergoes recurrent attachment and detachment from a message, as would be true of a ribosome subunit, a species of tRNA, or an initiation factor, then the rate of initiation would be slowed down relative to elongation, and Hb should be synthesized on very small polysomes, and possibly on monosomes. Conversely, if the average rate of elongation, or the availability of an initiation component which remains permanently associated with mRNA, limits translation of message beyond the saturation level, then the polysomes engaged in Hb synthesis should be as large (i.e. contain as many ribosomes) when message is above saturation as when it is below. Following this approach, Lingrel and Woodland (1974) have found that diagnostic peptides of both α and β globin can be detected on polysomes with 3–7 ribosomes whether mRNA is supplied above or below the saturating amount. This experiment therefore suggests that the translation of saturating amounts of injected Hb mRNAs is limited either by some aspect of the elongation process or by a limited supply of any component which is not used recurrently in protein synthesis. The involvement of a component which binds non-exchangeably with mRNA is suggested, as shown in Fig. 16 B.

The most direct way of investigating the lack of competition between injected and endogenous messages in oocytes is to carry out competition experiments with two different kinds of messages each of which is capable of saturating independently the translational capacity of oocytes. Under these conditions it would be possible to presaturate a cell with one kind of message, and then ask whether another kind of message can be translated, if introduced by a second injection. If oocytes have message-specific components (cf. Fig. 16A), the second message should be translated, even though a cell has been saturated with the first message. According to the other model (Fig. 16B), the second message would not be translated. To distinguish decisively between these two alternatives, two very different kinds of messages must be used in pure form, for example Hb mRNA and EMC virion RNA (see Table 9). EMC is used as an example of a message which is clearly different from Hb mRNA and is obtainable in very pure form, though the natural message to which it corresponds is not known. Compared to Hb mRNA, about 10 times the weight, and therefore about the same number of molecules, of EMC RNA is needed to saturate oocytes. This suggests that the component whose limited supply accounts for the saturation effect may be utilized in proportion to the number of mRNA molecules, and not, like ribosomes, to the number of nascent chains being formed. The competition results so far obtained have shown that once an oocyte has been saturated with one kind of message, the subsequent injection of the other type of message does not displace it or reduce its rate of translation (Laskey and Gurdon unpublished results). The ability of the oocyte to translate the second message is much reduced by having previously injected large amounts of the first. There is no proof from such work that message-specific factors exist, and all message-competition experiments so far carried out are consistent with the view that a messenger RNA molecule must first combine with a cell-component before it can be translated, and that this component is never dissociated from the message during translation (cf. Fig. 16B). The existence of such a component is sufficient to account for the observed saturation effects and lack of competition between messages.

All experiments so far completed fit the general concept of an unspecific message-activation molecule. There would be not more than roughly 10^{10} of these in each egg, assuming that one molecule can activate each message. Once a message had become associated with an activator molecule, it would be kept fully loaded with recycling ribosomes, and would remain stably associated with its activator. When all activator molecules (which would not distinguish different messages) had received a message, excess messages would not be translated at all, and might even be destroyed.

The results just outlined are still preliminary, but these experiments serve to show that living cells can be used satisfactorily for quantitative experiments of a precise kind. The fact that an oocyte can be injected with known volumes of material in two or more separate injections at any desired times over a period of many days, makes this cell-type exceptionally attractive for analysing the control of protein synthesis in living cells.

2.3.5. *The stability and intercellular distribution of messages*

Wide variation in the stability of different kinds of mRNA could have just as much effect on the kinds of proteins made in a cell as the control of transcription or translation of those messages. Although most bacterial messages are short-lived, having an average half-life of only a few minutes (Geiduschek and Haselkorn 1969), some animal messages are known to be relatively long lived. The best assessment of message half-life in animals comes from cells in which RNA synthesis terminates naturally, and in which protein synthesis, dependent on persistent mRNA, continues. This happens in some terminally specialized cells, such as mammalian red blood cells in which haemoglobin mRNA seems to have a half-life of 2–3 days (Marks and Kovach 1966). In cells which are synthesizing RNA, the stability of mRNA has to be determined by less direct methods. These involve the use of Actinomycin, which may affect the initiation of protein synthesis as well as transcription (Singer and Penman 1972), or the use of pulse-chase labelling which does not usually permit the life of one kind of message to be distinguished from that of another.

The insertion of mRNA into a cell which does not synthesize that type of mRNA permits a very direct test of the translational life of a message. (The translational life of a message— during which it continues to be translated—gives a minimum estimate of its actual life.) The following conclusions have been drawn from the injection of rabbit or mouse Hb mRNA into oocytes which were cultured for up to 2 weeks and labelled at various times during this period (Gurdon *et al.* 1974): (1) injected haemoglobin mRNA is about as stable as the average endogenous messages of the oocyte (Fig. 17). (2) The messages for mouse α and β globin are about equally stable in oocytes. (3) Within the cultural life of injected oocytes (which extends to 2 weeks), there is no detectable turnover of the injected messages. Although the efficiency of translation of injected messages decreases during prolonged culture, this can be fully accounted for by a decline in the efficiency of cultured oocytes' translational systems and does not imply that there is turnover of mRNA.

The fact that Hb mRNA is stable for well over a week in

oocytes makes it possible to ask whether the same messages are equally stable during development, if injected into fertilized eggs. A fertilized egg develops to the tadpole stage in about 5 days. Since development is characterized by the synthesis of new kinds of proteins, it is quite possible that a mechanism for destroying messages might exist in eggs and embryos, but not in oocytes, which synthesize the same kinds of proteins for several weeks. On the other hand, if injected Hb mRNA is as stable in embryos as it is in oocytes, this would at

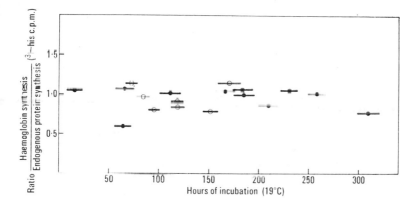

Fig. 17 Stability of haemoglobin mRNA injected into oocytes. Purified mRNA for mouse α and β globin was injected, at a subsaturating concentration, into oocytes which were then cultured in unlabelled medium. At intervals, samples were incubated in labelled medium for 12–24 hours (length of bar) after which they were frozen and analysed. The figure shows that there is no significant change in the ratio of haemoglobin to endogenous-protein synthesis for up to two weeks. ○, ●, oocytes from two different frogs (from Gurdon *et al.* 1974).

once open the way towards answering two important questions in development: does an egg possess a mechanism for distributing messages characteristic of one cell-type to only that part of an embryo? If not, can Hb mRNA be translated in cell-types such as muscle, which never normally synthesize haemoglobin?

This experiment is harder to achieve successfully than might be expected, because fertilized eggs are much more sensitive than oocytes to the disturbance of injected material; they often cleave abnormally, and much of the injected material is located in the uncleaved part of the embryo. Indeed Brachet *et al.* (1973), who were the first to publish the results of such an experiment, found that almost all injected eggs had become abnormal by the gastrula stage, after which haemoglobin synthesis was no longer seen in the Hb mRNA-injected embryos.

Recently, fertilized *Xenopus* eggs have been injected with mouse or rabbit Hb mRNA successfully, in such a way that several of the injected eggs developed entirely normally to the feeding tadpole stage (Gurdon *et al.* and Woodland *et al.* unpublished results). Rather surprisingly the synthesis of mouse or rabbit haemoglobin could still be detected in normal swimming tadpoles. In view of the enormous increase in the numbers of endogenous messages brought into use during development, the amount of haemoglobin synthesis *appears* to have decreased substantially between the egg and tadpole stages. However, allowance for this effect shows that injected Hb mRNA is about as stable in embryos as in oocytes. Tadpoles derived from message-injected eggs have been dissected into different regions, each of which is supplied independently with ^{3}H-amino acids. In each part, Hb synthesis is compared to the synthesis of endogenous proteins. It turns out that the region of the tadpole which contains the blood system synthesizes rabbit or mouse haemoglobin no more efficiently than other parts of the tadpole such as its muscle and vertebral column, which do not contain tadpole blood at this stage.

Two conclusions can be drawn from these experiments. First, it is clear that embryos possess no *general* mechanism for destroying all previously used messages at certain stages in development. Hb mRNA is rather stable both in oocytes and embryos. Therefore any messages which are turned over in development must owe their instability to a characteristic of their structure or location (making them especially sensitive to RNase), rather than to some general property of embryonic cells. It is already evident from experiments with Actinomycin on other species that two messages can have different half-lives in the same cell (e.g. Kafatos 1972). The other conclusion which emerges from the injection of mRNA into fertilized eggs is that no special mechanism exists in eggs whereby an injected message normally restricted to one cell-type can be distributed to only that part of an embryo; this does not of course exclude the possibility that messages synthesized naturally in early oogenesis may be distributed to particular regions of an egg. The overall effect of these experiments is to argue against the view that development and the early stages of cell differentiation are determined by the acquisition of message-specific components which ensure that a cell can translate messages of only the kind that it normally synthesizes. Presumably specialization of a cell's translational apparatus as suggested by Heywood's experiments (p. 49) confirms, rather than promotes, the differentiated state. In general, the ability to inject mRNA harmlessly into fertilized eggs which then develop normally greatly extends the usefulness of this method.

2.3.6. *Post-translational modification of proteins*

When a message has been translated, the newly synthesized proteins released from polysomes are not necessarily functional. Many proteins are secondarily modified, for example by scission of the primary protein chain, by elimination of amino acids from one end of the chain, or by phosphorylation, acetylation, etc. of the constituent amino acids, and changes of this kind may be essential for gene products to perform their function. This is the last step at which gene expression can be controlled, and it is clearly important to know whether developing cells become specialized in respect of their capacity to modify newly synthesized proteins.

Message injection experiments can provide information on the capacity of oocyte or egg cytoplasm to carry out secondary modifications to proteins not normally synthesized in these cells. The same information is not necessarily provided by *in vitro* extracts, since activities may be lost or gained by virtue of homogenizing cells and preparing extracts.

The ability of oocytes to modify proteins synthesized from injected messages is summarized in Table 10. It appears that all modifications characteristic of a protein are carried out in injected oocytes. It is conceivable that the enzymes needed to make these modifications are synthesized from mRNA present as a minor component in the message samples injected. In each case there are reasons why this is unlikely, as explained in the cited work, and in respect of collagen the presence of proto-collagen hydroxylase activity in eggs has been demonstrated in

TABLE 10

Modification of proteins synthesized in frog oocytes from injected mRNA

Protein	Cell-type from which mRNA is prepared and in which protein is synthesized	Normal protein modification	Protein modification in message-injected oocytes	Reference
αA2 crystallin	Calf lens epithelium	Acetylation of N-terminal methionine	N-acetyl-methionine in N-terminal peptide	Berns et al., (1972b)
Stable proteins of EMC virus	EMC virions from infected mouse ascites cells	Cleavage of primary polypeptide into mature proteins	At least six mature EMC proteins derived from different regions of primary polypeptide	Laskey et al., (1972)
Protamine	Trout testis	Phosphorylation of certain serine residues.	Phosphorylated peptides	Wu, Dixon, and Gurdon (unpublished)
Light chain immunoglobulin	Mouse myeloma cells	About 15 amino acids removed from end of primary polypeptide.	Mature light chains	Stevens and Williamson (1972) Smith et al. (1974)
Collagen	Cultured mammalian fibroblasts	Hydroxylation of proline in protocollagen.	Hydroxylation of ^{3}H-protocollagen (injected directly or synthesized from injected mRNA).†	Lane and Knowland (1974); Knowland et al. (1974).

† Since uninjected oocytes synthesize collagen, this test was carried out in unfertilized eggs which do not do so to a detectable extent.

extracts (Green *et al.* 1968), as well as in living cells by the injection of unhydroxylated protocollagen (Lane and Knowland 1974). These results suggest that protein-modifying enzymes capable of recognizing particular sequences of amino acids may be universally present in all cells during early development, and that the availability of these activities is not a step of which use is made to regulate gene expression during development.

2.4. The analysis of translation by '*in vivo*' and '*in vitro*' methods

Message injection into living cells has been specially emphasized in this chapter because it is a new technique which seems likely to become increasingly useful in the analysis of translational control, once the mechanism of protein synthesis is fully understood. Cell-free protein-synthesizing systems have been, and will continue to be, of primary importance in the purification of cell components, and in the identification of steps, involved in protein synthesis. On the other hand an understanding of the *control*, as opposed to the *mechanism*, of protein synthesis requires a knowledge of which step in the whole process is limited by an insufficient supply of a particular component. Cell-free systems can be so constituted that the rate of protein synthesis is limited by any desired component and this will not necessarily be the one which is limiting in the cell from which the system is derived. Variation in pH and in the concentration of different ions can substantially affect *in vitro* protein synthesis. Yet it is not known what pH, Mg^{2+} concentration, etc. exist in the region of a cell where protein synthesis is taking place. These uncertainties are avoided by injecting purified cell components directly into a living cell; the conclusions drawn will at least be valid for the conditions that exist in that cell-type. There is every reason to hope that, by combining the use of cell-free systems to analyze the mechanism of protein synthesis, and the use of injection into living cells to determine the role of each component in the control of protein synthesis, the importance of translation as a controlling step in gene expression during development may soon be understood.

2.5. Summary

There is reason to believe that translational control is responsible for the quantitative changes in protein synthesis which take place at fertilization in some species. The existence of a mechanism by which translational control can lead to changes in the kinds of proteins synthesized is exemplified by the effect of haemin on the efficiency of translation of α-globin mRNA in oocytes. Nevertheless it has been clearly extablished that oocytes and eggs have no mechanism by which they can exclude the translation of messages characteristic of other cell-types. Furthermore there is no concrete example of a substantial change in the kinds of proteins synthesized in development which can be certainly attributed to control of gene expression at the translational level. It seems likely that translational control is used to make quantitative adjustments to a pattern of protein synthesis determined primarily by the synthesis of new messages.

Special emphasis has been given in this chapter to the use of message-injection experiments which provide, perhaps uniquely, a means of analyzing experimentally the control, as opposed to the mechanism, of protein synthesis in living cells.

3 Gene transcription and the initiation of cell differentiation

3.1. Is gene transcription controlled ?

WE have concluded from the last two chapters, primarily by elimination, that transcription must be the level at which gene expression is primarily controlled. First, we must define the word transcription as used in this chapter. Gene transcription is the enzymatic synthesis of RNA from ribonucleotides which are assembled in a sequence complementary to one of the two strands of DNA. However the immediate product of transcription is probably not a functional message, but is thought to be a large molecule which is subsequently cut down in size (or 'processed') to yield the smaller, mature message which codes for protein (Penman *et al.* 1963; and Jelinek *et al.* 1973 for references to recent work). At some stage during or after this 'processing', mRNA is transported from the nucleus to the cytoplasm. Since no-one has yet identified the initially transcribed RNA from which a known message is subsequently derived, it is often convenient to recognize transcription by the formation of mature mRNA, and in this chapter, the term 'transcription' includes not only the synthesis of the initial precursor mRNA, but also the processing of this to mature mRNA and its transport to the cytoplasm. Is there any evidence that transcription is controlled?

3.1.1. *Polytene chromosomes of insects*

The cytological study of polytene chromosomes provides what is still one of the strongest cases in favour of transcriptional control in animals. There are good reasons for believing that the large 'puffs' or 'Balbiani rings', which are DNA-containing lateral expansions of multistranded interphase chromosomes, represent the sites of genes. In 1963 Beermann demonstrated in the midge, *Chironomus pallidivittatus*, that a relationship exists between a Balbiani ring located near the distal end of chromosome IV and the presence of granules in four special cells at the base of the salivary glands. In *Chironomus tentans*, on the other hand, neither the Balbiani ring nor the granular material is present in the special salivary gland cells. Hybrids between these two species show that granule production behaves as a single Mendelian factor, OR factor (gene), whose pattern of inheritance closely follows the location of the Balbiani ring on the IVth chromosome (Fig. 18). Recently Grossbach (1969) has shown that the salivary secretion of these two species differs in respect of a particular polypeptide. The presence of this polypeptide is related to the presence of a region at one end of chromosome IV where there is a Balbiani ring in *C. pallidivittatus*. This polypeptide is not synthesized in *C. tentans*. Balbiani rings show intense incorporation of ^{3}H-uridine into RNA, whereas the unpuffed bands do so to a much lesser extent (review by Berendes 1969). The distribution of puffs and Balbiani rings differs characteristically according to cell-type (Beermann 1956, 1973) and stage of development, especially in relation to the effects of the hormone ecdysone (Clever 1961). The combination of these relationships constitutes a strong case for the view that the Balbiani ring on the IVth chromosome represents the site of a gene or genes responsible for the accumulation of the polypeptide contained in the granular secretion of the special salivary gland cells. Transcription at this site is regulated in that the Balbiani ring is characteristically expanded (and active in RNA synthesis) or contracted at different stages of development.

The special advantage of polytene chromosomes is that RNA synthesis can be observed on enlarged chromosomes, and therefore at the site of transcription. This situation may be contrasted with other opportunities of recognizing transcription which often depend on the biochemical identification of mature messages, the formation of which may be affected by regulation at the levels of processing and transport. The only other type of cells, in which non-mitotic chromosomes have been studied

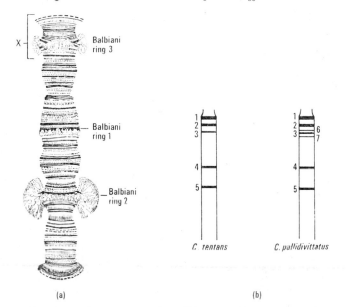

FIG. 18 Cytological evidence for differential gene transcription. The left hand figure (a) is a diagram of chromosome IV isolated from the salivary gland cells of the midge *Chironomus pallidivittatus*; transcriptionally active regions of the chromosome are called Balbiani rings. This diagram is based on a detailed cytological study of Beermann. The right-hand figure (b) shows acrylamide gel electrophoresis separations of protein sub-units extracted from the salivary gland secretion; band 7 is a polypeptide whose synthesis is correlated with the presence of genes at the distal end of chromosome IV where a special Balbiani ring occurs in chromosomes of the salivary gland, but not in chromosomes of other cell types. X marks the region of the chromosome in which the gene for the *C. pallidivittatus*-specific polypeptide is located. The equivalent chromosome of *C. tentans* does not possess a balbiani ring in the same region. (From Grossbach 1969.)

in such detail, are the oocytes of certain Amphibia (Callan 1963). The lampbrush chromosomes of these cells contain numerous densely-staining regions or chromomeres; some of these carry loops, which are believed to contain genes (Callan and Lloyd 1960), and which incorporate ³H-uridine into RNA very actively (Gall and Callan 1962). The majority of chromomeres do not carry loops, or synthesize RNA to a detectable level, and are presumed to contain transcriptionally inactive genes. The behaviour of lampbrush chromosomes makes it unlikely that transcriptional control at the chromosomal level is a unique phenomenon peculiar to specialized insect cells.

3.1.2. *RNA synthesis in amphibian development*

The clearest biochemical work which supports the existence of transcriptional control in development is that on well-defined direct gene products, such as 28s and 18s ribosomal RNA, 5s ribosomal RNA and 4s RNA; the synthesis of these has been studied in detail in the development of *Xenopus laevis*. Using the introduction of $^{32}PO_4$ into females just before ovulation, and the microinjection of ^3H-nucleosides at different stages of cleavage, to label embryos, it has been established (1) that 28s and 18s (though not 5s) rRNAs are always synthesized at the same relative rate (coordinately) during development, and (2) that 4s RNA, heterogeneous RNA (high molecular weight, DNA-like RNA other than rRNA and tRNA), 5s and 28s+18s rRNA are synthesized independently. The evidence for these statements comes from Brown and Littna (1964, 1966), Bachvarova and Davidson (1966), Gurdon and Woodland (1969), Landesman and Gross (1969), Abe and Yamana (1970). More recently, Ford (1971) and Mairy and Denis (1972) have shown that 5s ribosomal RNA is synthesized in great excess of 28s and 18s rRNA during the early stages of oogenesis of *Xenopus laevis*, and this provides another example of non-coordinate synthesis of different classes of RNA. It is known that the 5s RNA synthesized in early oogenesis and the 5s RNA synthesized in development are coded for by different genes (Ford and Southern 1973). Figure 19 summarizes the principal changes in the pattern of synthesis of different nucleic acids during *Xenopus* development.

The generalization that the major classes of RNA are synthesized at independent rates during early development, and this is clearly true of *Xenopus*, appears also to be valid for all other animal species tested; see, for example, Schwartz (1970) for *Urechis*, and Sconzo and Giudice (1971) and Emerson and Humphreys (1971) for sea urchins, Woodland and Graham (1969) and Knowland and Graham (1972) for mice; experiments on several other species are referred to in reviews by Brown (1965) and Gurdon (1968c).

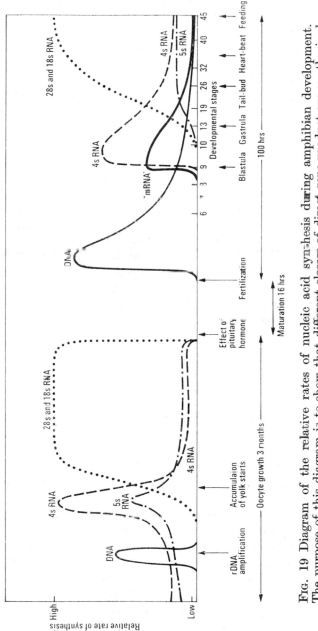

Fig. 19 Diagram of the relative rates of nucleic acid synthesis during amphibian development. The purpose of this diagram is to show that different classes of direct gene product are synthesized at non-coordinated rates during oogenesis and development. Rates of synthesis are indicated very approximately on a low to high scale. The diagram indicates, very roughly, the proportion of an oocyte's or embryo's total nucleic acid synthesis which is constituted by DNA or by a class of RNA. This diagram is based on the results of papers referred to on p. 78; 'mRNA' = heterogeneous RNA (p. 78).

A similar generalization cannot be made about another aspect of transcriptional control, and this concerns the stages in development at which different kinds of gene transcription commence. In several animal species, nuclear RNA synthesis cannot be detected at all during the first few cleavage divisions. In view of the small number of nuclei per embryo at these stages, autoradiography provides a more sensitive test than biochemical analysis for this conclusion, which applies for example to *Xenopus* (Gurdon and Woodland 1969), the mollusc *Ilyanassa* (Collier 1966), and, among insects, to the beetle *Leptinotarsa* (Lockshin 1966) and the milkweed bug *Oncopeltus* (Harris and Forrest 1967). However, in mammals (Mintz 1964) and sea urchins (Nemer 1963; Kedes and Gross 1969), new RNA synthesis can be detected already at the two-cell stage. It is impossible, in such cases as *Xenopus*, to prove that no RNA is synthesized at all, since it might be turned over or transported to the cytoplasm much faster than normal, but there does appear to be a real deficiency of gene transcription in the early embryos of some species. This question is mainly of importance in connection with the rate of ribosomal RNA synthesis, which, in contrast to other classes of RNA increases from gastrulation onwards more or less in relation to cell number. For this reason it can be argued that rRNA synthesis could easily be over-looked during the first few hours of development (Emerson and Humphreys 1970, 1971). However even when this is taken into account, Sconzo *et al.* (1970) and Sconzo and Giudice (1971) find an activation of rRNA synthesis in *Paracentrotus*, a different genus of sea urchin from *Strongylocentrotus*, that studied by Emerson and Humphreys. Furthermore, Landesman (1972) has shown that the difference in rate of rRNA synthesis in pre- and post-gastrula stages of *Xenopus*, cannot be accounted for solely by the 10-fold difference in cell number.

The kinds of RNA of most interest in development are those which code for proteins, and these may be regulated in quite a different way from genes whose final product is RNA, such as ribosomal or transfer RNA. As yet, there is no entirely satisfactory way of measuring the synthesis of individual messages during development. Numerous studies have compared heterogeneous populations of RNA molecules by hybridization to total cellular DNA, with either RNA or DNA limiting the reaction. The sensitivity of hybridization reactions can be enhanced by observing the hybridization at different Cot values (e.g. Davidson and Hough 1971), and it may soon be possible to use these methods to determine accurately the numbers of RNA molecules which code for *identified* proteins. It can be shown that specialized tissues synthesize large amounts of mRNA for the kind of protein which they characteristically produce. For example, a close relation exists between the amount of ovalbumin mRNA and the rate of ovalbumin synthesis in chicken oviduct cells (Schimke *et al.* 1973); a similar conclusion has been reached by Suzuki *et al.* (1972) in respect of silk fibroin. However, embryos do not synthesize very large amounts of any one kind of protein and mRNA synthesis in embryos has not been examined with the same precision. But when comparisons are made by less precise molecular hybridization procedures, differences are always observed in the populations of RNA which have been accumulated or are being synthesized at different developmental stages (e.g. Denis 1966; Whiteley *et al.* 1966). In conclusion, biochemical experiments have shown that mRNA synthesis is controlled in specialized cells, and suggest that this is also true of embryos.

3.1.3. *The nucleolus and gene expression*

It has been suggested that, although the synthesis of ribosomal and messenger RNAs may be independent, the passage of mRNA from the nucleus to the cytoplasm may be dependent on the presence of a nucleolus and hence on rRNA synthesis (Harris *et al.* 1969). The synthesis of rRNA could in this way regulate mRNA transport to the cytoplasm and hence gene expression. Harris and his colleagues (1969) indeed observed a temporal relationship, under a variety of experimental conditions, between the appearance of a nucleolus (which is related in development to the onset of ribosomal RNA synthesis), and the initiation of new gene expression, after a transcriptionally inert erythrocyte nucleus has been fused into a growing cultured cell. On the other hand the anucleolate mutant of *Xenopus laevis* is able to reach the swimming tadpole stage with functional muscle, blood, lens cells, etc. in the total absence of ribosomal RNA synthesis (see Fig. 6, for details and references). There are two simple ways in which this apparent discordance may eventually be resolved. The first is that the nucleolar characteristic which is linked to gene expression in fused cultured cells may be unconnected with ribosomal RNA synthesis; that is, the passage of mRNA from nucleus to cytoplasm may be dependent on the presence of a nucleolus, but not necessarily on ribosomal RNA synthesis. The second explanation assumes that cultured cells and developing embryos differ in the following way. In a rapidly growing line of cultured cells, it is very likely that ribosomes are synthesized at just the rate needed to service the amount of mRNA which is made; therefore additional ribosome synthesis would always be required for new message translation when a nucleus is fused into a cultured cell. In *Xenopus*, it has been demonstrated directly that spare ribosomes exist in the egg and are available for the translation of new messages (Chapter 2); new ribosomes and rRNA synthesis are not therefore required to translate additional messages during early development. Whatever the ultimate explanation of the results with fused cells, the fortunate existence of a mutant from which all ribosomal genes have been deleted (details in Fig. 6) makes it clear that, in development, there is no necessary connection between 28s and 18s rRNA synthesis and new gene expression (including message transport); new gene expression, but not rRNA synthesis, occurs several days before such tadpoles die.

We can conclude that different classes of genes are regulated independently at the transcriptional level during development, as shown most clearly by the different patterns of synthesis of ribosomal and 4s RNAs. It is also clear that there is no necessary connection in amphibian development between ribosomal RNA synthesis and the expression of mRNA genes.

3.2. Is the control of gene transcription important in development ?

3.2.1. *The effect of inhibiting RNA synthesis*

The fact that populations of heterogeneous RNA change substantially in early development does not itself prove that the genes which code for protein are transcribed at this stage. There are however several reasons for believing that this is the case. One is that agents such as Actinomycin or X-irradiation, which inhibit RNA synthesis but which have little immediate effect on protein synthesis, arrest development within a few hours when applied to embryos of sea urchins (Gross 1967), frogs (Brachet and Denis 1963), molluscs (examples in Collier 1966), fish (Neyfakh 1971), and mice (Mintz 1964), etc. These effects suggest an essential role of transcription in development, but do not prove this because these agents may cause developmental arrest for other reasons than the inhibition of RNA synthesis.

3.2.2. *Paternal gene expression*

In hybrids between strains or species whose genes code for distinguishable proteins, the appearance of a paternally specified gene product proves that transcription must have taken place between fertilization and the time of its appearance; the appearance of this geno product could not have resulted from the translation of stored messages, and messages introduced with sperm should be translated immediately after fertilization. The activity of paternally specified enzymes such as the dehydrogenases of 6-phosphogluconate and lactate have been detected at the muscular response stage of frog embryos (Wright and Moyer 1966), at the heart-beat stage of quail embryos (Ohno *et al.* 1968), and already at the blastocyst stage in mice (Chapman *et al.* 1971). It is however likely from the early lethality of interspecies hybrids that new gene transcription in non-mammalian species takes place long before these relatively advanced embryonic stages. Moore has shown that diploid hybrids between various species of *Rana*, such as *R. sylvatica* egg and *R. pipiens* sperm, are arrested at the late blastula or early gastrula stage, but that haploids within the species (such as *pipiens* sperm and enucleate *pipiens* egg) develop to the tail bud stage (review by Moore 1955). Therefore the presence of an extra chromosome set of a foreign species causes very early developmental arrest at the blastula stage, and this occurs before any gross chromosome deficiencies arise. Since nuclear RNA synthesis is first detected at the mid-late blastula stage in amphibia (see above), hybrid developmental arrest may well result from the wrong kind of gene activation by the foreign cytoplasm. In *Drosophila* there are some very early acting lethal mutants of a non-maternal type. Embryos homozygous for the mutation *deep orange* (*dor*) are arrested at the stage of primary organogenesis (Hildreth and Lucchesi 1967), (Fig. 20), and those homozygous for Notch-8 show

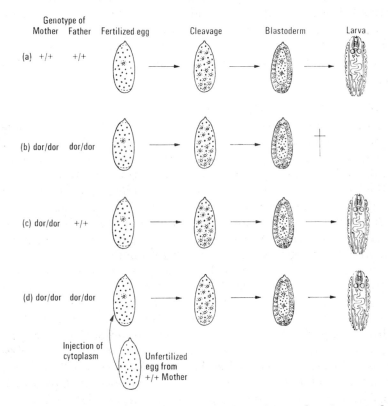

FIG. 20 *Drosophila* development: genetic and cytoplasmic control. (a) normal development. (b) development arrested during gastrulation through the maternal effect of the recessive mutation deep orange (*dor*). (c), (d) the deficient development of *dor/dor* eggs can be corrected by wild-type sperm (c) or by the injection of cytoplasm from wild-type eggs (d).

abnormal differentiation of the ectoderm at the blastoderm stage (Poulson 1945). It is possible that some genes on the X-chromosome need to be expressed before the blastoderm stage, since embryos totally deficient for X-chromosomes

(Nullo-X of Poulson 1945) display an abnormal distribution of cleavage nuclei. Abnormal development does not take place when an X-chromosome-deficient egg is fertilized by an X-chromosome-containing sperm. Other examples of mutants which may act very early in development are reviewed by Wright (1970).

In conclusion, new gene expression seems to have commenced before the late blastula (or blastoderm) stage in those species which have been most fully investigated. In some species new gene activity may take place soon after fertilization, but in others such as amphibia this seems not to be the case.

3.3. The regional distribution of materials in fertilized eggs

We now turn to the problem of how to investigate the control of gene transcription, and in a broader sense the control of nuclear activity. RNA can be synthesized *in vitro* from chromatin, and from nucleo-protein, as well as from purified DNA if RNA polymerase is added. However it is still uncertain how accurately genes can be transcribed *in vitro*. For example it has yet to be shown that the ribosomal DNA of *Xenopus*, about which more is known than any other plant or animal genes (Brown 1967; Birnstiel *et al*. 1971) can be transcribed *in vitro* with added polymerase in such a way that only the correct DNA strand is read, and the spacer region (see Fig. 2) is not transcribed. Another difficulty in this field is that although it seems clear that cells possess a special nucleolar (ribosomal DNA) RNA polymerase (Roeder and Rutter 1970), this enzyme does not show any preference for ribosomal DNA over bulk chromosomal DNA (Roeder *et al*. 1972). Difficulties of this kind have impeded progress in the study of transcription *in vitro*. Partly for this reason and partly for the reasons stressed in Chapter 2 (p. 73) there is a strong incentive to look for conditions in which the control of nuclear activity can be studied in living cells, and this approach is emphasized in the rest of this chapter.

One of the oldest, and probably still the most important, 'principles' of animal development is that a relationship exists between particular regions of a fertilized egg and certain types of differentiation in a developing embryo. The validity of this relationship must be appreciated if current attempts to analyse the control of gene activity in living cells are to be understood.

3.3.1. *Cytoplasmic localizations of importance in development*

A classic demonstration of this point is provided by Conklin's description of the normal development of ascidian embryos belonging to the genus *Styela*. Early embryos at the 64-cell stage have five clearly demarcated regions of cytoplasm, recognizable by virtue of natural differences in colour. Cytoplasm of each kind is destined to be confined to cells of a certain type, such as coelomic mesoderm, notochord, endoderm, etc. These different kinds of cytoplasm cannot be distinguished in unfertilized eggs, but some regions such as the yellow crescent (later contained in mesoderm cells) becomes clearly evident a few minutes after fertilization. Although most animal embryos are not so fortuitously provided with regionally-specific coloured cytoplasm, the same general conclusion is valid, that certain regions of egg or early embryo cytoplasm are related to certain types of differentiation. This has been established for many different species, by tracing the lineage of specialized cells back to very early blastomeres, as well as by killing or separating certain early blastomeres and observing that the undamaged or isolated blastomeres develop the same structures as they would have in an intact embryo. One of the most spectacular examples of the latter kind is the removal of the polar lobe of cytoplasm from the two-cell (or 'trefoil') stage of the early mollusc embryos (*Ilyanassa* or *Dentalium*; Wilson 1904). Experiments of these kinds have been reviewed in detail by Wilson (1925), and by Davidson (1968). Although most of the examples which establish a connection between regions of egg cytoplasm and certain types of differentiation have come from invertebrates, this relationship is also true for vertebrates. In amphibia, a 'grey crescent' is formed in a meridional region of the uncleaved egg opposite the point of entry of the sperm, and this region constitutes the site of the blastopore lip at gastrulation; this locates the future dorso-ventral axis of the embryo, the anterior–posterior axis having been previously determined by the distribution of yolk. Local damage to the grey crescent cytoplasm of an uncleaved egg, but not damage to any other region of the egg, prevents axis formation (Curtis 1962; Brachet and Hubert 1972). A second point of gastrulation and a second axis can be induced in the ventral part of an embryo if a small piece of cytoplasmic cortex is grafted from the grey crescent of one egg to the ventral region of another (Curtis 1962).

These comments would be deficient without mention of what is probably the most outstanding example of a cytoplasmic localization, applicable to both vertebrates and invertebrates, the germ-plasm or pole-plasm. In many Dipteran insects, one end of the elongated egg contains a histologically distinct region of cytoplasm, called the pole plasm, into which two nuclei migrate at the 16-nucleus stage. The mitotic products of these nuclei constitute the pole cells which later give rise to sperm or eggs of male or female flies. A variety of ingenious experiments, involving the displacement or destruction of the pole plasm, by centrifugation, constriction, or u.v. irradiation, have established a close relationship between the pole plasm and germ-cell differentiation (Geigy 1931; Bantock 1970; Geyer-Duszyǹska 1959), so that eggs without pole plasm develop into flies which are normal except for being sterile. A similar situation exists in amphibian eggs, and here too destruction of cytoplasm at the vegetal pole of the egg by u.v. light leads to the formation of animals which are normal in all respects except that they are wholly sterile (Bounoure *et al.* 1954; Blackler 1962; Smith 1966). The importance of all these examples is that they demonstrate a close relationship between a particular region of egg cytoplasm and a defined type of cell differentiation. In each case, a region of egg cytoplasm evidently contains molecules or components which cause the cells with which they become associated to differentiate in the way described.

The process exemplified by these cases is one in which an egg contains a heterogeneous distribution of 'morphogenetic' materials. These are gradually laid down during the growth of the oocyte (oogenesis), and a further rearrangement usually takes place at fertilization. The result is a large cell which contains many different regions of cytoplasm, each destined to promote a certain type of differentiation. During cleavage, many identical nuclei are rapidly formed and these come to occupy different regions of the egg cytoplasm. Early cleavage divisions are always very rapid, and therefore a few hours after fertilization, each nucleus is surrounded by a different kind of cytoplasm, which is thought to activate or repress genes, and so lead to the first initial differences between the cells of an embryo. A fertilized egg is to be thought of as a closed system, in which a redistribution of, and synthesis of molecules from, materials already present in the egg accounts for the transition of a single fertilized egg cell to a complex embryo and larva consisting of many different specialized cells. This general principle of development was recognized at the end of the nineteenth century and remains a cornerstone of embryology.

3.3.2. *Cytoplasmic distribution in adult plants and animals*

Is the importance of unequal cytoplasmic distribution limited to early development? Cell differentiation continues throughout life both in plants and in animals, as for example, in the replacement of skin, intestine, and blood cells, etc. It now seems that the unequal distribution of cytoplasm from parent to daughter cells may well be a common feature of cell differentiation in adult organisms, as well as in embryos. Grasshopper neuroblasts, for example, undergo a series of divisions at each of which one of the two daughter cells differentiates as a fully specialized ganglion cell and the other as another neuroblast; only the latter divides again to repeat the process. In the dividing neuroblast, the metaphase spindle is consistently orientated so that it is predictable which end of it will give rise to the neuroblast or ganglion daughter cells. Carlson (1952) was able to insert a very fine needle and rotate the metaphase spindle so that those chromosomes which would have migrated into the ganglion daughter cell entered instead the neuroblast daughter cell, and vice versa. In spite of this, the parts of the parent cell cytoplasm destined to form a ganglion or neuroblast cell did so. This is an elegant demonstration of the fact that it is regional differences in parent cell cytoplasm rather than any characteristics of the daughter chromosomes that determine the different directions of daughter cell specialization. The same type of situation may well be true of many other kinds of dividing adult cells in animals, such as skin.

The rigid walls and vacuoles of plant cells make them very hard to manipulate in the way that can be done for animal cells. However the unequal division of cells commonly observed in plant differentiation suggests that the same generalization may be true of plants. An example of this is stomatal differentiation in the epidermal cells of the grasses (Fig. 21).

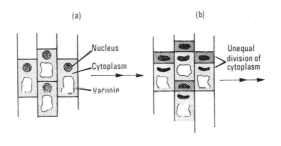

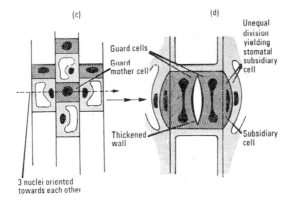

FIG. 21 An unequal distribution of cytoplasm in dividing plant cells, correlated with cell differentiation. The diagram shows dividing epidermal cells in a grass leaf. In (a) and (b), four cells undergo unequal cytoplasmic division, one of each pair of daughter cells being vacuolated. In (c), the nuclei of the vacuolated daughter cells become orientated towards the nuclei of non-vacuolated cells. In (d), the vacuolated daughter cells undergo another unequal cytoplasmic division, after which the non-vacuolated daughter cell becomes a stomatal subsidiary cell. At the same time, the guard mother cell undergoes an equal division into the two stomatal guard cells, whose nuclei become dumb-bell shaped. The guard cells acquire thickened walls on each side of the stomatal opening. (From Clowes and Juniper 1969.)

3.4. Opportunities to analyze the effect of egg cytoplasm on development

There are two approaches to this problem. In one, the aim is to 'rescue' an egg whose development is genetically or experimentally deficient by injecting cytoplasmic material from another egg capable of developing normally. If this can be done it becomes possible, in principle, to identify the active component, by fractionating the cytoplasm and testing each fraction for its curative effect. The other approach is to select a well-defined nuclear response to cytoplasm, and attempt to purify from the cytoplasm a component which causes it. The first type of experiment can take advantage of genetic mutants, and so ensure that the cytoplasmic component being purified is a developmentally important gene-product. It suffers from the disadvantage that the biological effect used to recognize the 'rescue' is usually rather poorly defined, such as a stage of early development reached, and this makes it very hard to analyze the mode of action of the purified cytoplasmic component. The second approach has the advantage of dealing throughout with a very clearly defined effect, the molecular details of which are relatively well understood. Its disadvantage is that it is not always easy to design a biological assay capable of proving that the purified component is in fact the one that operates in normal development. Experiments of the latter kind were reviewed by Gurdon and Woodland (1968).

3.4.1. *The rescue of developmentally deficient embryos by microinjection of cytoplasm*

Rescue experiments designed to analyse cytoplasmic effects were first used successfully by Smith (1966) in animal embryos. Fertilized frogs' eggs were irradiated with u.v. light at the vegetal pole, where the germ plasm lies close to the surface, so that all the resulting tadpoles lacked germ-cells but were otherwise normal. Material from the vegetal pole of another non-irradiated egg was then sucked out and injected into a set of u.v.-irradiated eggs, and a significant proportion (25–50 per cent) of these then developed with germ-cells, though not with as many as are found in non-irradiated controls. The u.v. action spectrum of germ-cell inactivation suggested that the sensitive component is nucleic acid and not protein. Unfortunately it is not easy to pursue this attractive type of rescue experiment, partly because only small amounts of germ-plasm can be conveniently obtained, and partly because the biological response (germ-cell formation) is completely undefined biochemically.

Another very elegant type of rescue experiment, designed by Briggs, is more easily amenable to analysis, and in this case use is made of eggs incapable of normal development owing to a maternally-inherited genetic defect (Briggs and Justus 1968; Briggs 1969). As indicated in Fig. 22, eggs of o/o females are arrested early in development if fertilized by wild-type sperm, and even if injected with a wild-type blastula nucleus. However if injected with cytoplasm from a wild-type egg, development of o/o eggs often proceeds nearly normally. It turns out that the curative material is highly concentrated in the nucleus of the oocyte, and this makes it easy to prepare very large amounts of concentrated material for fractionation and assay. The active material must be the direct or indirect product of a single gene, and probably (though not necessarily) of one which is active in the oocyte itself. Even in this case the low viability of mutant animals and the time-consuming nature of the assay system limit the rate of progress. It seems likely that the curative agent is a large molecule (judged by sedimentation rate), and a protein (judged by trypsin sensitivity). Other 'maternal lethal' mutants have been collected in the axolotl ('f', 'v' of Briggs 1969) as well as in another Urodele, *Pleurodeles* ('ac' of Beetschen and Jaylet 1965; and Beetschen 1970). However such mutants can be collected much more easily in *Drosophila*, and a rescue experiment has been carried out on *deep orange*, using a stock in which *dor/dor* embryos die during gastrulation (Garen and Gehring 1972). About one quarter of the embryos injected with egg cytoplasm from *wild-type* eggs develop to the more advanced stage of muscular movement (Fig. 20). The identification of the corrective component is likely to be difficult because the small size of *Drosophila* eggs makes them hard to inject harmlessly, and the volume of fluid which can be injected is about 100 times less than that tolerated by frog embryos.

It is clear from these comments that rescue experiments can

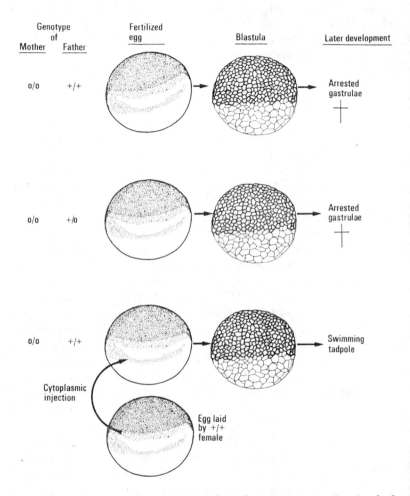

FIG. 22 Correction of a maternal lethal mutation in the Axolotl. A recessive gene o, which stands for 'ova deficient', not for gene deficiency, is active during oogenesis in the Axolotl. Its product is necessary for eggs laid by a female homozygous for the o-gene to develop beyond the blastula stage (second row). Eggs of homozygous females will not develop beyond the blastula stage, even if fertilized by wild-type sperm (upper row). However the injection of cytoplasm from a wild-type oocyte or egg permits development to proceed up to the tadpole stage (lower row). (From Briggs and Justus 1968).

be successfully performed; it is unfortunate that amphibia whose eggs are most amenable for such experiments are so inconvenient for genetic purposes on account of their long life-cycle. However as more comes to be known about the probable nature of the effective cytoplasmic components, rescue experiments are likely to become increasingly useful. At present it is still a matter of pure speculation whether the agents which correct maternal lethals do so by contributing a messenger RNA (perhaps stabilized by associated protein), a protein which activates genes, or by some other means.

3.4.2. *Effects of oocyte or egg cytoplasm on nuclear activity*

We turn now to circumstances in which egg cytoplasm has a well-defined effect on nuclear or gene activity. One kind of work which falls into this category is that on chromosome diminution and elimination in *Ascaris* (Boveri 1899) and *cecidomyid* flies (referred to above). Nuclei which enter the pole plasm are delayed in division and also spared the loss of chromosomes which is undergone by all other (somatic) nuclei. The difficulty in trying to analyse this kind of situation is that two different regions of the same egg cytoplasm have opposite effects, and it is not yet possible to separate large amounts of cytoplasm of the two kinds, nor to test the effect of components purified from such eggs. It is furthermore questionable whether the mechanism which leads to chromosome loss will give any information about normal mechanisms of gene control.

Two conditions are known in which the cytoplasmic environment of a nucleus can be altered experimentally in such a way as to induce a substantial change in gene activity: nuclear transplantation and cell fusion. In each case, nuclei are induced to undergo changes which are normal, as opposed to pathological, and the whole, as opposed to only one part, of the recipient cell cytoplasm has the inducing effect. The last characteristic makes it possible to try to extract, and analyse biochemically, the inducing activity from recipient cell cytoplasm, an approach which is not feasible where two parts of the recipient cell cytoplasm have opposing effects, as in the cases of chromosome elimination and germ-plasm.

Nuclear transfer experiments. The magnitude of changes

induced by nuclear transplantation is illustrated particularly clearly by the results of transplanting skin cell nuclei. There is no known way of converting an intestine cell or committed skin cell into blood, muscle, lens cells, etc. Yet the mitotic products (daughter nuclei) of successfully transplanted nuclei promote differentiation of these radically different types (as explained on pp. 18–25). Clearly nuclear transplantation leads to the activation of haemoglobin, myosin, and crystallin genes in intestine or skin cell nuclei, and the changes are presumably induced by a normal mechanism since they take place in the course of normal development.

To investigate the mechanism by which nuclear transplantation causes changes in gene expression, two series of experiments have been carried out, both of which demonstrate in egg cytoplasm the existence of components which have a rapid and well-defined effect on nuclear activity. In the first experiment (Graham *et al.* 1966; Gurdon 1968*b*) multiple adult frog brain nuclei, which synthesize DNA extremely rarely, were injected into three types of recipient cell (Fig. 23): (*a*) ovarian oocytes, which synthesize RNA, and especially rRNA, but no DNA; (*b*) ovulated oocytes which are undergoing the completion of meiosis and whose chromosomes are condensed in the course of meiotic division; and (*c*) unfertilized eggs, whose nucleus synthesizes DNA but not RNA, immediately after activation (or nuclear injection). In each case adult brain nuclei from the same sample were induced within a few hours to change activity or function so as to conform to that characteristic of the cell by whose cytoplasm they were then surrounded (Fig. 23). This experiment demonstrates clearly the ability of egg cytoplasm to induce nuclei to undergo changes which are normal for the host cell. Of special value for analytical purposes is the fact that oocytes and eggs, two cell-types readily obtainable in large numbers, have opposite effects on easily recognizable activities (DNA and RNA synthesis) of injected nuclei. In overall composition, oocytes and eggs are very similar, one being convertible by hormone action within a few hours to the other (Fig. 3), and yet they differ fundamentally in respect of a cytoplasmic component which influences normal nuclear activity.

The second type of nuclear transfer experiment which

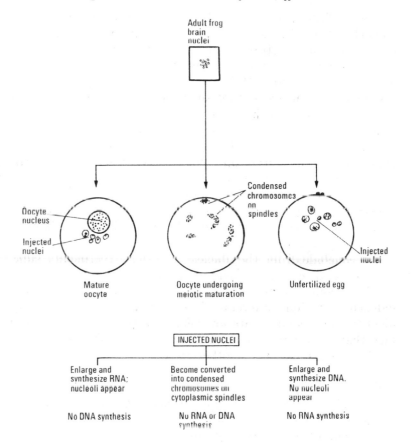

Adult frog
brain
nuclei

Oocyte
nucleus

Injected
nuclei

Condensed
chromosomes
on
spindles

Injected
nuclei

Mature
oocyte

Oocyte undergoing
meiotic maturation

Unfertilized egg

INJECTED NUCLEI

Enlarge and
synthesize RNA;
nucleoli appear

No DNA synthesis

Become converted
into condensed
chromosomes on
cytoplasmic spindles

No RNA or DNA
synthesis

Enlarge and
synthesize DNA.
No nucleoli
appear

No RNA synthesis

Fig. 23 Cytoplasmic control of nuclear activity. Adult brain nuclei, which do not normally synthesize DNA or pass through mitosis, are made to change activity in different ways by exposure to different kinds of oocyte or egg cytoplasm. The responses shown take place within one hour in eggs, within 4 hours in maturing oocytes, and within 24 hours in a mature oocyte.

demonstrates the importance of cytoplasmic control involves the transplantation of a single neurula nucleus to an enucleated egg (Gurdon and Woodland 1969). As explained above (p. 78 and Fig. 19), there is a natural sequence of changes in the pattern of RNA synthesis during the first 15 hours of development. If a neurula nucleus which is actively engaged in the synthesis of all major classes of RNA is transplanted to an egg, all nuclear RNA synthesis ceases within 20 minutes. As development of the nuclear-transplant embryo proceeds, each class of RNA is seen to be synthesized according to the normal sequence of events (Fig. 24 b).

Components of egg cytoplasm have therefore been able to reversibly repress each main class of RNA synthesis. This cannot be accounted for in any simple way, by for example the deficiency of active RNA polymerase, since the transplantation of a nucleus of one species into the enucleated egg of another results in renewed synthesis of some but not other classes of RNA (Woodland and Gurdon 1969); Fig. 24c). Apparently egg cytoplasm contains different components capable of regulating independently the activity of different classes of genes.

We conclude from the nuclear transfer experiments summarized that egg cytoplasm contains an inducer of DNA synthesis and a repressor of RNA synthesis which operate independently on different classes of genes. Attempts to identify these components are very greatly facilitated by the fact that they are absent or inactive in oocytes which are similar to eggs in so many other respects.

Cell fusion experiments. The type of cell fusion experiment of special interest for the present purposes is one in which genes not normally active in a cell are induced to become so, as a result of an altered cytoplasmic environment. In most cell-hybrids, the differentiated types of gene expression are lost, but in some cases specialized characteristics are retained (Ephrussi 1973). In one case the induction of new kinds of gene expression has been described. Among the progeny of hybrids between 3T3 mouse fibroblasts and tetraploid rat hepatoma cells, Peterson and Weiss (1972) isolated five clones three of which synthesized mouse albumin, even though 3T3 fibroblasts do not normally do so. This experiment demonstrates that it is possible to activate those genes which are never normally expressed in an adult cell, by fusion, just as was described above for nuclear transplantation. In each case several cell generations elapse between the time of fusion or nuclear transfer and the demonstration of new gene expression.

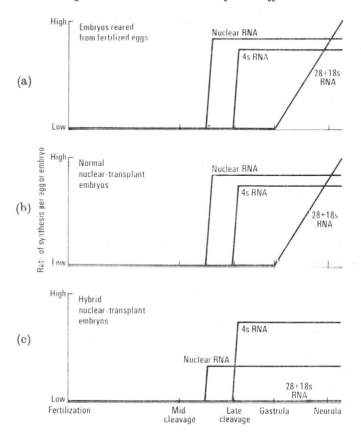

Fig. 24 Cytoplasmic control of RNA synthesis in amphibian development. Enbryos were injected with ³H-uridine or ³H-guanos ino and incubated for 2 hours before the extraction and analysis of labelled RNA. The pattern of RNA synthesis is shown under three conditions. (a) normal development of fertilized eggs (*Xenopus laevis*); (b) normal development of enucleated unfertilized eggs which received single transplanted neurula nuclei; (c) development of enucleated unfertilized eggs of *Discoglossus pictus*, a Sardinian frog in the same family but not genus as *Xenopus*. 'Nuclear RNA' = heterogeneous RNA (p. 78). (From Gurdon and Woodland 1969; Woodland and Gurdon 1969.)

To investigate the mechanism of gene activation after fusion it is necessary, as with nuclear transfers, to examine more immediate nuclear responses. This has been done by Harris *et al.* (1969) who have studied gene expression in hen erythrocyte nuclei fused into cultured mammalian cells, before the first post-fusion mitosis takes place. Little if any cytoplasm is carried over with the erythrocyte nucleus, which is in effect placed in the cultured cells' cytoplasm. Mature hen erythrocytes are inactive in RNA synthesis, but about 4 days after fusion into a cultured cell active in RNA synthesis, some of its genes are reactivated, such as those responsible for chick-specific surface antigens, inosinic acid pyrophosphorylase activity, and diphtheria toxin sensitivity (Harris *et al.* 1969; Deák *et al.* 1972). Not all genes are reactivated (Hb is not synthesized) and those which are were all previously active in the immature erythrocyte. Nevertheless this experiment clearly demonstrates the ability of cell cytoplasm to regulate the activity of genes responsible for the synthesis of known proteins.

The experiments outlined in this section have provided direct evidence that cytoplasmic components regulate gene activity in early development, and perhaps in adult cells as well. These experiments therefore substantiate the importance which embryologists have for many years been attaching more and more firmly to egg cytoplasm. Above all, they open the way for attempts to work out the molecular basis of the effects of cytoplasm, since they clearly imply that eggs contain molecules which can control defined types of nuclear or gene activity.

3.5. The nature and mechanism of cytoplasmic control

It is not obvious that either DNA synthesis or rRNA synthesis have an immediate effect on the kinds of proteins synthesized in early development. The reason for devoting special attention to these activities is that the induction of DNA synthesis and the repression of rRNA synthesis both exemplify easily recognizable, immediate effects of egg cytoplasm on nuclear activity. Both are normal effects of egg cytoplasm on the male and female pronuclei which exist in eggs immediately after fertilization. It is essential to try to analyse, first, the simplest examples of an effect of egg cytoplasm on nuclear activity, and turn only later to the much more difficult problem of how genes which code for proteins are regulated.

3.5.1. *DNA synthesis*

The cytoplasmic induction of DNA synthesis can be regarded as an interaction between two very complex components: egg cytoplasm and a nucleus. The *in vivo* analysis of this interaction would be enormously simplified if a nucleus could be replaced by purified DNA, thereby eliminating any essential role for the poorly defined non-DNA components of the nucleus. It now seems clear that purified DNA will serve as a template for DNA synthesis if injected into unfertilized eggs, and that this reaction represents the normal response of nuclear DNA to egg cytoplasm. The reasons for this claim, which is greatly strengthened by the results with polyoma virus DNA, are the following. (*a*) The injection of purified DNA of any kind into unfertilized eggs causes a substantial incorporation of ³H thymidine into DNA, as does the injection of whole nuclei (Gurdon *et al.* 1969). There appears to be a 20-minute lag between the time of injecting double-stranded (native) DNA or nuclei and the onset of DNA synthesis, as is also observed between the time of fertilization and the onset of DNA synthesis in gamete pronuclei. (*b*) The injection of native DNA or whole nuclei into oocytes does not stimulate DNA synthesis (Gurdon and Speight 1969). Single-stranded (denatured) DNA is replicated in both oocytes and eggs, and fails to show the 20-minute lag period observed in eggs (Gurdon and Speight 1969). Two kinds of bacterial DNA failed to stimulate DNA synthesis when injected into eggs in double stranded form, but did so if first denatured (Gurdon *et al.* 1969). Evidently native vertebrate DNA must be used to reveal cytoplasmic effects characteristic of normal eggs and oocytes. (*c*) The kind of DNA synthesized is similar to the DNA injected as judged by buoyant density analysis of ribosomal DNA (Gurdon *et al.* 1969). A very sensitive test of this is provided by injecting closed circles of supercoiled native polyoma virus DNA, as a result of which daughter replicas, also closed, supercoiled, and native, are synthesized (Laskey and Gurdon 1973; Fig. 25). The same experiments indicate semi-conservative replication of the injected polyoma DNA. (*d*) As the amount of injected DNA is increased, the amount of DNA synthesis which is stimulated constitutes a decreasing proportion of that injected (Gurdon

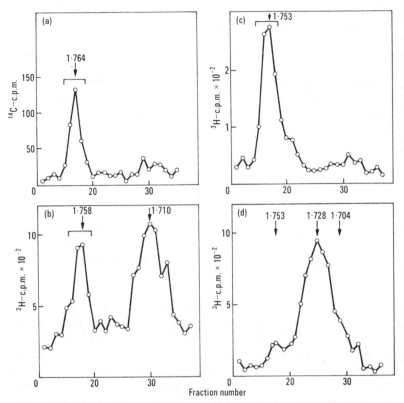

FIG. 25 Evidence for replication of polyoma virus DNA in frog egg cytoplasm (from Laskey and Gurdon 1973). Polyoma infected cells were allowed to incorporate the density label BUdR for a limited amount of time, so as to label only one of the two polyoma DNA strands. The resulting closed circles of native polyoma DNA, in a supercoiled configuration, had a buoyant density of 1·764 g/ml in CsCl (Fig. 25a). (The DNA was also lightly labelled with ^{14}C to facilitate purification). DNA from the bracketed region of Fig. 25a was collected and injected into eggs, together with ^{3}H-thymidine. DNA was than extracted from the injected eggs and analysed on a CsCl gradient, with the result shown in Fig. 25b. Part of the ^{3}H-labelled DNA had a buoyant density 1·71 g/ml, as expected of supercoiled closed circles which lack a density label in either DNA strand. The other part of the labelled DNA had the density of part-heavy part-light supercoils (1·758 g/ml). This is the result expected if the parent light: heavy polyoma DNA had replicated, adding a light daughter strand to each parental strand. The bracketed region of Fig. 25b was then collected and repurified on another CsCl gradient (Fig. 25c). This material was then denatured to separate the two DNA strands, and it was then shown, by another CsCl gradient (Fig. 25d), that almost all the ^{3}H-thymidine label was contained in the light (1·728 g/ml) strand. These are the results expected if the parental polyoma DNA was replicated semi-conservatively in the frog egg cytoplasm.

et al. 1969). When an amount of polyoma DNA equivalent to *ca.* 1000 nuclei is injected, the newly synthesized polyoma DNA amounted to 18 per cent of this (Laskey and Gurdon 1973), and would be higher for smaller amounts of injected DNA. The conclusion from these experiments is that purified native DNA of vertebrates appears to be replicated when injected into the living cytoplasm of eggs but not of oocytes. Egg cytoplasm must therefore contain all enzymes, etc. required to replicate DNA. This greatly simplifies attempts to identify the component of egg cytoplasm which normally induces DNA synthesis after fertilization, since the inducing factor is absent from oocytes and can be investigated independently of the many poorly-understood non-DNA components of a nucleus.

A simple way of accounting for the ability of egg cytoplasm to induce DNA synthesis is to suppose that oocyte cytoplasm lacks the necessary DNA precursors or DNA polymerase Woodland and Pestell (1972) found that, of the common DNA precursors, dCTP and dTTP are present in oocytes in substantial amounts, but dATP and dGTP could not be detected, the limit of detection being the amount needed to synthesize about 200 nuclei.

Two very direct tests of the importance of DNA precursor supply can be carried out on oocytes (Woodland *et al.* 1971). One involves the injection of DNA precursors so as to create in an oocyte at least as high a concentration of each common DNA precursor as exists in eggs. The other involves the injection of denatured DNA which stimulates DNA synthesis in oocytes. It was shown that most of each injected sample of deoxyribose triphosphate remains as a triphosphate in the oocyte, and that a high intra-cellular concentration of all precursors failed to induce DNA synthesis when whole nuclei or native DNA are injected into oocytes. The amount of DNA synthesis stimulated by the injection of denatured DNA was enough for the replication of more than 100 nuclei. We can therefore say with confidence that although the supply of DNA precursors increases as oocytes are converted to eggs, this is not the reason why egg cytoplasm (but not oocyte cytoplasm) induces DNA synthesis. The fact that the concentration of precursors in a living cell can be substantially raised, and that DNA which stimulates incorporation of precursors can be injected, makes a negative conclusion of this kind much firmer than is normally possible.

Two studies have been made of DNA polymerase activities that can be extracted from *Xenopus* oocytes and eggs. Grippo and Lo Scavo (1972) found that unfertilized eggs contain two DNA polymerase activities (assayed with denatured DNA as a template), which are distinguishable by their pH optima as well as by their isoelectric points. The enzyme with an iso-electric point at pH 7·0 is not found in oocytes. Benbow and Ford (unpublished work) have extracted from eggs a DNA polymerase activity which shows a preference for a nicked native DNA template; oocytes do not contain an activity of this kind. In principle, the appearance of a new kind of DNA polymerase during oocyte maturation might account for the induction of DNA synthesis by egg cytoplasm, especially since the oocyte chromosomes are in a condensed condition (and therefore presumably unresponsive to DNA polymerase) throughout the maturation period, as discussed by Gurdon (1967*a*). It would however be rather surprising if this were the whole explanation for the difference in the nuclear activity of oocytes and eggs. Fortunately this possibility can be tested directly when the egg polymerase has been purified, by injecting it into oocytes.

It is worth mentioning briefly some attempts to reproduce *in vitro* the induction of DNA synthesis by egg cytoplasm. Barry and Merriam (1972) have used *Xenopus* egg cytoplasm to induce erythrocyte nuclei to undergo a substantial swelling, an event which immediately precedes the induction of DNA synthesis in whole eggs (Fig. 23; Graham *et al.* 1966). The only claims to have initiated DNA synthesis *in vitro* with cyto-plasmic extracts have resulted from the addition of cytoplasm from rapidly growing cultured cells (such as HeLa) to nuclei from cultured cells or erythrocytes; a modest stimulation of DNA synthesis was observed (Thompson and McCarthy 1968; Mueller 1969; Kumar and Friedman 1972). Fractionation of the active cytoplasmic component has revealed two factors, differing in heat stability, in the two sets of experiments referred to. The principle shortcoming of such *in vitro* systems is that the amount of DNA synthesis initiated *in vitro* is very small, averaging, apparently, less than 0·1 per cent of a complete replication. The stimulation of DNA synthesis in these systems is not yet sufficient to permit definite conclusions to be drawn about the nature of the cytoplasmic factors concerned.

3.5.2. *Ribosomal RNA synthesis.* The demonstration that rRNA synthesis is reversibly repressed by egg cytoplasm (Gurdon and Brown 1965; Fig. 24) soon led to attempts to extract an inhibitor of rRNA synthesis from eggs or cleaving embryos before the stage of detectable rRNA synthesis. Shiokawa and Yamana (1967) found that isolated neurula cells, which synthesize rRNA rapidly, were largely inhibited from doing so by incubating them with isolated mid blastula cells, or with the medium in which blastula cells have been previously incubated; if neurula cells were incubated with neurula cells instead of blastula cells, rRNA synthesis was not inhibited. The inhibitor was reasonably selective since it did not affect the synthesis of 4s RNA, DNA, or heterogeneous RNA. Under the most effective conditions, the ratio of rRNA : 4s RNA synthesis was reduced from 6 : 1 (in control neurula cells) to 2 : 1 (neurula cells in inhibitory medium). A reason for believing that this inhibitor is normally involved in the regulation of rRNA synthesis is the close relationship that exists between the stages of development from which the inhibitor can be extracted and those at which rRNA is not synthesized: an impressive example of this is the fact rRNA is synthesized in the animal but not vegetal half of gastrulae (Woodland and Gurdon 1968) and that the inhibitor can be extracted from the vegetal but not animal half of such embryos (Wada *et al.* 1968). These results suggest that a natural inhibitor of rRNA synthesis is released from dissociated blastula cells. The inhibitor is probably a small molecule and not a protein, since it is dialysable, resistant to boiling for 10 minutes, and is not sedimented at 100 000 g for 2 hours (Shiokawa and Yamana 1967).

Several attempts to repeat these results were unsuccessful (e.g. Landesman and Gross 1968; and Hill and McConkey 1972), possibly because the conditions of cell dissociation may be critical for achieving the release of the inhibitory factor. However Laskey *et al.* (1973) have recently succeeded in preparing a 0·5 N-perchloric acid extract of eggs which is capable of totally inhibiting rRNA synthesis when added to dissociated neurula cells, so that rRNA synthesis is reduced from its normal level (several times more than 4s RNA synthesis) to an undetectable amount (Fig. 26). Laskey *et al.* were unable to detect any selective inhibition of rRNA synthesis when neurula cells were incubated with blastula cells or with the medium in which they

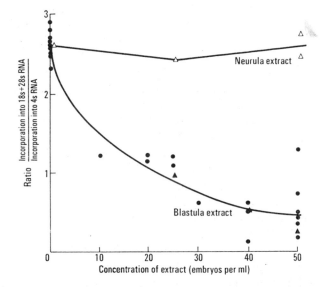

FIG. 26 A cytoplasmic inhibitor of ribosomal RNA synthesis. Eggs or embryos of *Xenopus laevis* are extracted with perchloric acid. The extract is partially purified by absorbtion to charcoal, and then added to isolated cells of neurula embryos. If the extract is prepared from blastulae which do not synthesize ribosomal RNA, it inhibits ribosomal RNA synthesis in the neurula cells (●, ▲, results with extracts from two unrelated samples of blastulae). If the extract is prepared from neurulae, which normally synthesize ribosomal RNA, it fails to inhibit ribosomal RNA synthesis (△). Each point represents the result of an electrophoretic analysis of ³H-uridine labelled RNA extracted from the cells of 5 neurulae (from Laskey *et al.* 1973).

had been maintained. Nevertheless, it is clear that early *Xenopus* embryos do contain an inhibitor of rRNA synthesis, and this may well be the component responsible for the reversible inhibition of rRNA synthesis in transplanted nuclei.

Crippa (1970) has attempted in a different way to isolate an inhibitor of rRNA synthesis, using full-sized *Xenopus* oocytes which remain in an ovary 3 days after hormone-induced ovulation; oocytes of this kind are usually active in rRNA synthesis (e.g. Gurdon 1967*a*), but were presumably not so in the experiments to be outlined. Isolated germinal vesicles (nuclei) from these oocytes were subjected to homogenization, ammonium sulphate precipitation, and fractionation on DNA-cellulose and then DEAE–sephadex columns; the fractions obtained were then tested for activity by injection into immature oocytes. Until the last step of the procedure, all fractions had a generally inhibitory effect, but finally one fraction was obtained that inhibited rRNA but not 4s RNA synthesis in immature oocytes which were allowed to incorporate ³H-uridine for 2 hours after injection. The inhibitory fraction, believed to be a protein, bound to ribosomal DNA, but not to chromosomal DNA free of rDNA. In view of the surprising source and complex procedure by which this inhibitor was isolated, it would be reassuring if this interesting result could be confirmed, preferably with material from unfertilized eggs which never synthesize rRNA at a detectable rate.

Quite a different approach to the regulation of ribosomal RNA synthesis is provided by a range of mutants which constitute deletions for various numbers of ribosomal RNA genes. The first of these to be discovered in *Xenopus*, the Oxford O-*nu* mutant, involves a total deletion of ribosomal RNA genes (see Fig. 6). Individuals heterozygous for the mutation (1-*nu*) are as viable as wild-type animals but have only half the number of rRNA genes per diploid nucleus. In spite of this, 1-*nu* and wild-type (2-*nu*) tadpoles synthesize rRNA at the same rate (Brown and Gurdon 1964; Gurdon and Woodland 1970). Dosage compensation of this type is very unusual; it is not observed except in the case of antibody-coding genes, and genes on the X-chromosome of species with XX:XY sex chromosomes. If the number of copies of a gene responsible for an enzyme activity is increased (without a corresponding increase in the rest of the genome,) it is generally observed that the amount of that enzyme activity is proportionately increased. Even in cases where one dose of a gene gives the same phenotype as two doses (e.g. the gene R for round seed in peas: Mendel 1866), it may turn out, as in this case, that the amount of enzyme activity concerned (in this case for conversion of sugar to starch) is proportional to the number of these genes per genome. It appears that the activity of genes which code for proteins, and which are probably present in only one copy per chromosome set, may in general be limited by the amount of template present (see also p. 113). In contrast, the behaviour of 1-*nu* and 2-*nu* *Xenopus* tadpoles implies that the amount of template does not limit the activity of the multiple rRNA genes; only when the number of rRNA genes per diploid nucleus is reduced to below about ⅓ of the wild-type number, is rRNA synthesized at a subnormal rate and only then are the embryos or tadpoles inviable (Miller and Gurdon 1970: Knowland and Miller 1970).

rRNA gene mutants have been collected in substantial numbers in *Drosophila* (through their '*bobbed*' phenotype— review by Ritossa *et al.* 1966) and in moderate numbers in *Xenopus* (through the associated nucleolar condition), (Miller and Gurdon 1970). In no case has a compelling argument been made for the existence of any kind of regulatory mutants. It therefore seems that, at present, the simplest way of accounting for the regulation of rRNA genes in development is to suppose that these genes are always available for transcription (de-repressed), but that the rate or frequency of their transcription is governed by the activity of a specific rRNA polymerase; this in turn would be governed by the concentration of a low molecular weight inhibitor. The conditions which would lead to the presence of the inhibitor are quite unknown. However the observed facts, other than the activity of amplified ribosomal genes in oocytes, could be explained if the inhibitor were synthesized at the same rate as ribosomes but sequestered by ribosomes which are incorporated into polysomes.

3.6. The passage of proteins from cytoplasm to nucleus

It is a reasonable assumption that the influence of cytoplasm on nuclear activity is mediated either by the removal of molecules from the nucleus or by the passage of cytoplasmic molecules into the nucleus. The early effects of cytoplasm on the nucleus seem not to be accompanied by a substantial loss of nuclear proteins. This was tested by transplanting nuclei

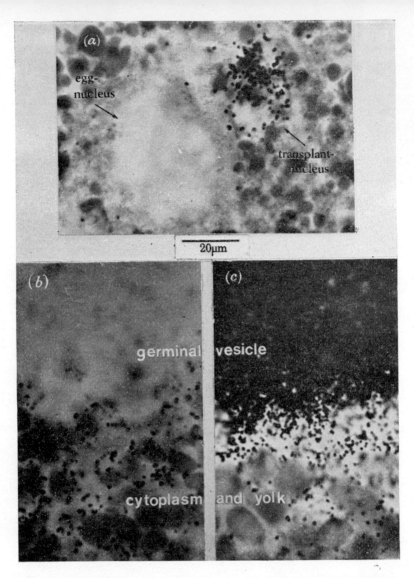

FIG. 27 Exchange of proteins between nucleus and cytoplasm in oocytes and eggs. (From Gurdon 1970). (a) A test for the passage of proteins from a transplanted nucleus to egg cytoplasm. An embryo was labelled from the blastula to the neurula stage with mixed ^{3}H-amino acids. Nuclei, now containing labelled proteins, were then transplanted to eggs which were fixed for autoradiography 50 minutes later. There was no obvious reduction in the label of the transplanted nucleus, and no labelled proteins were transferred to the adjacent egg nucleus, which had not been U.V.-irradiated in this experiment. (b), (c) Selective uptake of cytoplasmic proteins by oocyte nuclei. Iodine^{-125} was chemically attached to purified bovine serum albumin and calf thymus histones, which were then injected into oocytes. Autoradiography of the sectioned oocytes showed that within 3 hours, the labelled histones had become strongly concentrated in the nucleus (Fig. 27c), but the bovine serum albumin was still much more concentrated in the cytoplasm.

containing labelled proteins to eggs. Autoradiography of the eggs one hour after nuclear transfer (when a change in activity had been induced) showed that the transplanted nuclei were still heavily labelled (Gurdon 1970; Ecker and Smith 1971). A more sensitive test for the loss of nuclear proteins is obtained by looking for radioactivity in the egg nucleus, which can be allowed to move next to the transplanted nucleus; no transfer of labelled proteins of the kind observed by Goldstein and Prescott (1967) in *Amoeba* was seen in *Xenopus* (Gurdon 1970; and Fig. 27a).

The passage of proteins from egg cytoplasm to a transplanted nucleus was first demonstrated by injecting brain nuclei into unfertilized *Xenopus* eggs whose cytoplasm had previously been labelled with ^{3}H-amino acids, and which were supplied with puromycin at an intracellular concentration sufficient to reduce protein synthesis to less than 2 per cent of normal (Arms 1968; Merriam 1969). Autoradiography revealed that within 90 minutes labelled proteins had become strongly concentrated in the injected nuclei to several times the cytoplasmic level. The accumulation of proteins in injected nuclei is temporarily related to the induction of DNA synthesis, and the small number of nuclei which fail to synthesize DNA also fail to accumulate protein (Merriam 1969). In comparable experiments with *Rana pipiens*, Ecker and Smith (1971) showed that cytoplasmic proteins synthesized during oocyte maturation are strongly concentrated in nuclei even at the late blastula stage. In cell-fusion experiments in which an erythrocyte nucleus is induced by cultured cell cytoplasm to resume DNA and RNA synthesis, Bolund *et al.* (1969) observed a two-fold increase in nuclear dry mass 1–2 days after fusion. All these results show that a general correlation exists between the cytoplasmic induction of new nuclear activity and the passage of proteins from the cytoplasm to the nucleus. Clearly the possibility of a causal relationship exists.

To examine this relationship in more detail, it is essential to know what kind of proteins pass from the cytoplasm to the nucleus. Microinjection makes it possible to follow the fate, and determine the intracellular function, of *single kinds* of proteins which have been purified, identified, and labelled *in vitro* with ^{125}I (Gurdon 1970). Since the specific activity of the ^{125}I protein sample is known before injection, the amount of radio-activity in the nucleus (determined autoradiographically or by dissection) gives an accurate estimate of the number of molecules of the injected protein in the nucleus. In the first experiments of this kind, which were carried out on oocytes, it was found that histones become highly concentrated in the germinal vesicle, and that apparently normal living oocytes can be obtained with at least 100 times more histone in their nuclei than is normally present on their chromosomes (Fig. 27, b,c; Gurdon 1970). To determine the function of histones on nuclear activity, the same number of molecules, now in unlabelled form, can be injected into oocytes which are then labelled with ^{3}H-uridine and analysed autoradiographically or biochemically. This experimental procedure offers, perhaps uniquely, a means of determining the function of a known concentration of a known kind of protein on chromosome activity of living cells. The procedure will be of greatest use if it can be applied to those amphibian species with oocytes which have a small number of large 'lampbrush' chromosomes and which are resistant to microinjection; it should then be possible to determine the chromosomal sites to which proteins bind, and their effect on individual loops or groups of genes.

The first experiments with injected iodinated proteins have been extended by Bonner (unpublished), who has examined the behaviour of several proteins differing in size and charge. The nuclear membrane presents a strong, but not complete, barrier to the passage of molecules of molecular weight above 60 000 which enter an interphase nucleus very slowly, whereas smaller proteins reach equilibrium much more quickly (within 1–2 days). It is however possible that some naturally synthesized oocyte proteins may behave differently, since some molecules of more than 60 000 in molecular weight can enter the nucleus. The pronounced accumulation of histones in the nucleus is not seen with other proteins of similar size and isoelectric point, such as lysozyme.

The purpose of this section has been to emphasize the probable importance of cytoplasmic proteins in regulating nuclear activity, and to describe an experimental procedure which should prove valuable in attempts to determine the behaviour and function of nuclear proteins.

3.7. Developmental agents of extracellular origin

We have implied that cytoplasmic control is of primary importance in development and have argued that the unequal distribution of materials in eggs must account for some of the differences between cells. It is likely, however, that unequal cytoplasmic distribution initiates, but does not fully account for, cell differentiation. It is well known that some degree of 'regulation' is possible in many kinds of animal eggs, since isolated blastomeres from the two-cell stage, and in some cases from the 8-cell stage, can form complete embryos (e.g. Waddington 1956; Davidson 1968). It would, furthermore, be difficult to understand how an uncleaved egg could possess enough precisely-localized regions to account for all cell-types in a larva. Probably differences between cells are initiated by the unequal distribution of cytoplasm, and then amplified by extracellular agents, a term which includes cells, materials released by cells, and physical properties of the environment. Some examples of extracellular agents which affect early development are now outlined.

One of these occurs in embryonic induction, a term which refers to those occasions during development when one cell layer is influenced by another cell layer which has come to lie next to it as a result of movements of cell layers in development; these movements occur continually throughout development, from gastrulation (primary induction) onwards. Although the details of embryonic induction are not yet well understood, one important characteristic has emerged. This is that the specificity of the induction reaction depends to a large extent on the responding tissue which must be 'competent' to react to the inducing tissue (see Yamada 1967*b*, and Tiedemann 1967). The inducing tissue, or substances released by it, may be considered to determine the stage in development when the responding tissue reacts and the exact amount of the tissue which responds. The nature of the response, and the fact that one tissue and not another is capable of responding, depends on properties of the competent tissue. Thus a tissue which is induced in development has already differentiated in respect of its ability to respond to an inducer.

Another type of extracellular agent of importance in development is a hormone. Hormones are certainly involved, for example, in the many changes that take place during metamorphosis in insects and amphibia (e.g. Weber 1967; Tata 1971), and in sex differentiation. However the same point of view as was expressed for embryonic inducers also applies to hormones; these can affect only those cells which have, in the course of their differentiation, acquired the appropriate receptors. We conclude that both hormones and inducers are important in controlling the timing of specialization during development. Both agents, however, depend on cells having acquired molecules which enable them to respond. The distribution of molecules which enable a cell to respond to an inducer need not be very accurately defined in embryos, since the geographic limits of induction may be determined by the exact area of contact between two-cell layers.

Another category of extracellular agent may be involved in bringing about the dichotomous differentiation of daughter cells in such cases as the grasshopper neuroblasts (p. 88) and plant leaf epidermal cells (p. 89). In such cases it would not be surprising if environmental agents cause components of a parent cell to be unequally distributed to its daughters. We have seen that an unequal distribution of cytoplasm is closely related to the divergent differentiation of daughter cells. This type of effect may take place whenever a monolayer of stem cells, with a different environment on each side, yields daughter cells which behave divergently, as for example in mammalian skin cell differentiation.

Finally a quite different type of developmentally-important extracellular agent is a sperm cell. A very important consequence of fertilization is the resulting rearrangement of materials in an egg, thereby imposing a dorso-ventral axis on a fertilized egg, which before fertilization possessed only an anterior–posterior axis. While sperm penetration may not universally have this effect, it certainly does so in several groups of animals (e.g. grey crescent formation in amphibia).

The aim of this section has been to point out that, while extracellular agents are certainly important in animal development, many examples of such agents may have their effects by altering the distribution of intracellular regulatory molecules or by interacting with receptors already contained in or on a cell; in either case these agents would supplement or enhance a pre-existing mechanism which involves intracellular regulatory molecules, and would not carry information specifically related to the response which they elicit.

3.8. Summary

The main points to emerge from this chapter are the following. Gene transcription is controlled in development. Egg cytoplasm contains a range of components which are of major importance in development because they regulate independently the activities of nuclei and genes, and because they appear to be responsible for promoting different directions of specialization in early development. Attempts to identify developmentally important cytoplasmic components have included (1) the injection of cytoplasm into eggs whose development is genetically or experimentally defective, and (2) the fractionation of egg cytoplasm which has clear-cut effects on the activity of transplanted nuclei. Separate components which induce DNA synthesis and which repress ribosomal RNA synthesis are present in the cytoplasm of eggs but not of oocytes. Some classes of genes (such as the multiple copy ribosomal genes) appear to be regulated in a different way from others (such as single-copy protein-coding genes). Extracellular agents which have an effect on development may do so solely by influencing the distribution of cytoplasmic materials.

Special emphasis has been placed on experimental procedures which may permit the identification and functional analysis of cytoplasmic molecules which regulate nuclear activity and gene expression.

4 *General Conclusions*

THE following comments constitute an amplified summary of some of the points of view expressed in Chapters 1–3. These are brought together into a speculative concept of how gene expression might be regulated in development. We discuss very briefly the level at which gene expression is regulated, the limiting steps of transcription and translation, the distinction between cell-type specific and universal genes, and lastly the relationship of cell division to the regulation of gene expression.

Concerning the level at which gene expression is regulated in development, we have given reasons for excluding changes in the kinds or numbers of genes (Chapter 1), and post-translational modification (Chapter 2, p. 71), as steps at which regulation generally takes place. Some cells certainly contain message-specific components which could affect gene expression at the level of protein synthesis, but the distribution of these in eggs and specializing cells suggests that they play a supplementary role in accentuating, rather than initiating, differences between cell-types (Chapter 2, p. 70). Transcription (including the processing and transport of messages) is probably the level at which gene expression is principally controlled in development (Chapter 3).

The next important question concerns the step(s) at which translation and transcription are limited. Translation is discussed first. If we knew that the amount of protein synthesized by a cell was limited by the supply of initiation factors, we would concentrate attention on the specificity and synthesis of these factors rather than on the conditions which govern the production of other translational components. It is possible

that the limiting component for protein synthesis will turn out
to differ according to the particular protein made, the cell-type,
and the stage of development. Nevertheless, in most cases
where it is possible to guess what the limiting component is, it
seems likely to be the supply of mRNA. This has been demonstrated conclusively in oocytes and eggs of *Xenopus*; here the
supply of extra message results in a *proportionate* increase in the
amount of message product synthesized (Chapter 2). In agree-
ment with this, it is most unlikely that ribosomes limit protein
synthesis in development. In all species examined, fertilized eggs
contain a very large number of ribosomes, only a small proportion of which are engaged in protein synthesis. Only when
early development is complete have most of the egg's ribosomes
become incorporated into polysomes. Furthermore there is no
case in development where the supply of initiation factors,
tRNA, etc., has been clearly shown to limit the rate of protein
synthesis. We therefore assume that protein synthesis is
usually limited in development by the supply of mRNAs, and
hence by regulation at the transcriptional level.

At the level of transcription, it seems useful to regard genes
as falling into two classes whose transcription is regulated in
different ways:

1. Genes which are potentially active in all cell-types, which
are present in multiple copies in each genome, and whose
transcription is limited in some other way than by template
availability (e.g. by supply of active RNA polymerase).
2. Genes which are expressed in only a limited range of cell-
types, which are present in only one copy per genome, and
whose transcription is limited by template availability.

The distinction between these two classes of genes has been
proposed in one form or another by various authors and is
suggested by certain characteristics of cell differentiation.
Genes of the first category are sometimes referred to as 'house-
keeping genes', since they provide the essentials for cell main-
tenance and proliferation, in contrast to 'cell-type specific'
genes of group (2) which synthesize 'luxury' products of
principal value to the organism as a whole and not to the cell
in which they are synthesized. In single-celled organisms, where
the cell *is* the complete organism, almost all genes will fall into

the first category; all such genes are potentially active in every cell. In contrast, multicellular organisms possess large numbers of specialized cells the majority of whose total protein synthesis is coded for by type (2) genes.

The justification for placing all genes into the two categories, (1) and (2), is tenuous, but the distinction seems valid in the following examples. The numbers of genes per haploid *Xenopus* genome has been shown to be multiple for ribosomal RNA (500 genes), 5s rRNA ($\sim$25 000 genes), 4s RNA (1000 genes), and histones (100 genes), but is almost certainly single or nearly so for rabbit α and β globin, silk-moth fibroin, and hen ovalbumin (Chapter 1). Products of genes of the former group are synthesized in almost all kinds of cells; products of the other (single-copy genes) are synthesized to a detectable level in only a small percentage of an animal's cells, and these genes appear to be totally inactive in over 90 per cent of all cells. Single-copy genes are generally transcribed in a dose-dependent way, so that heterozygotes synthesize half as much gene product as homozygotes. The only obvious exceptions to this rule are genes in mammalian X-chromosomes and antibody-forming genes. In contrast, a two or more fold reduction in the numbers of ribosomal RNA genes (as in one-nucleolus mutants of *Xenopus* and 'bobbed' mutants of *Drosophila*) leads to no detectable reduction in the amount of ribosomal RNA synthesis (Chapter 3). This last comparison suggests that the transcription of single-copy genes is limited by the amount of template (genes) available. Since this is not so for multiple-copy genes, we assume that genes of this class are never masked but are subject to regulation in all cells, so that the amount of product synthesized fits the metabolic requirements of the cell, as must generally be true for bacterial genes. This could be achieved, in the case of animal ribosomal RNA, by the interaction of a type of RNA polymerase and an inhibitor as suggested on p. 106.

If single-copy, cell-type specific, genes are regulated by the availability of template, we must suppose the existence of gene-specific molecules which interact with such genes. Acidic nuclear proteins are in principle capable of fulfilling this role (Paul 1970), and could act in early development as follows. They would be synthesized continually by the genes which code for them, and distributed unequally in a fertilized egg.

As a result, different kinds of proteins in an egg would be distributed to nuclei in different regions of a blastula, so that blastula cells are made to synthesize different messages and hence different proteins. The consequent differences between cells will be poorly defined at first, but will be amplified and more precisely delineated as a result of embryonic folding movements and inductions (p. 109). The same kinds of cytoplasmic proteins would be unequally distributed between daughter cells in later stages of development (pp. 88 and 89).

An additional point worth mentioning concerns the relationship of the supposed cytoplasmic regulatory molecules and the genes whose activity they govern. Cell division has commonly been associated with changes in gene expression, and there are some indirect arguments in favour of the idea that cell-type specific genes are regulated only once per cell cycle, a point of view examined in detail elsewhere (Gurdon and Woodland 1970). Reprogramming of genes at each cell cycle could be understood if the proteins which regulate genes are released from, and recombined with, chromosomes at the beginning and end of mitosis, as has been observed by the use of antibodies synthesized by individuals with autoimmune diseases (Beck 1962; Ringertz *et al.* 1971). In general a daughter cell would be expected to reprogram its chromosomes for the same kinds of activities as those of its parent cell. But this would not be true whenever the cytoplasm of a daughter cell is different from that of the parent cell. Such a difference could arise not only during cleavage when nuclei populate heterogeneous egg cytoplasm, but also as a result of the influence of the extracellular agents described on p. 109-111.

In conclusion, development is visualized as a series of interactions between a nucleus and a changing cytoplasmic environment. It is hoped that the methods outlined in Chapters 1-3 will be useful in elucidating the nature of these interactions.

Appendix

A. The cloning of animals

THE vegetative propagation of plants is a routine procedure and many genetic strains of horticultural or agricultural value are maintained only by this means. Many species of lower invertebrates have remarkable powers of regeneration, but among the vertebrates there is no species in which a complete adult animal can be propagated from part of another adult, the remaining part surviving independently. In only very few animal species can more than one complete organism be prepared from an *early embryo* beyond the two-cell stage, and in most species this can be done only by separating the first two blastomeres. Yet it is desirable to be able to produce clones of genetically identical vertebrates, partly because these would be useful for certain types of immunological or behavioural experiments where individual variation needs to be kept to a minimum, and partly because it could be useful to propagate certain individual animals endowed with a particularly felicitous genetic constitution.

Genetically-identical adult vertebrates which are normal and fertile were first produced in 1961 by transplanting nuclei from an embryo into a number of enucleated recipient eggs of *Xenopus* (Gurdon 1961a). In some experiments as many as six serial transfer generations (explained in Chapter 1 and Fig. 5) were made, and clones containing over 30 genetically-identical adults were prepared. There would be no difficulty in making a clone of several hundred frogs by this means. The frogs within one clone all carried a nuclear marker (Fig. 5), were all of the same sex, and all had a similar skin pattern (Gurdon 1961*a*). In another series of experiments, a clone of frogs, produced by transplanting nuclei from one subspecies into egg cytoplasm of another, all had the characteristics of the 'nuclear' and not 'cytoplasmic' subspecies (Gurdon 1961*b*). Skin grafts between members of a clone are accepted permanently, but are rejected within 1 to 2 weeks when exchanged between siblings of a normal mating (Gurdon 1964). Within the *Xenopus* clones referred to, some animals were undersized, but a similar proportion of undersized animals is commonly obtained when the progeny of a normal mating is reared in the laboratory under the same conditions. These examples show that it is possible to make clones of vertebrate animals which are genetically identical as judged by the available criteria.

We have discussed so far the production of clones from cells of an early embryo, but the frequency with which nuclear-transplant embryos develop normally decreases when cells from advanced embryos or larvae are used as nuclear donors (pp. 28). It is however desirable to be able to propagate an adult animal whose characteristics are already apparent, and it would be even more useful if cells from an animal could be frozen, stored, and then used successfully for nuclear transplantation. It has been found that nuclear transfers from freshly isolated adult cells give very poor results, but success is enormously improved if cells which grow out from an adult tissue are cultured for a few days and then used as nuclear donors (pp. 22). Normal adult frogs have been prepared, by serial transplantation, from the nuclei of cells which have been in culture for several weeks (Gurdon and Laskey 1970), but which originated from tadpole epidermis. The best development so far obtained from cells of an adult animal is the recently metamorphosed frog shown in Fig. 7 b. The ability to transplant nuclei successfully from a cultured cell means that there should be no difficulty in transplanting nuclei from frozen stocks of cells, and it is doubtful if there is a real difference in transplantability between cells grown from larval as opposed to adult epidermis. It is therefore likely that an adult frog could be prepared from a cell of another adult. It is not yet known whether such an animal would be fertile, though there is no obvious reason why it should not be. It must however be appreciated that the preparation of an adult frog from a cultured cell nucleus is technically very difficult and so far the method by which this can be done has been successful only in amphibia.

The possibility of carrying out nuclear transplantation in mammals has stimulated a lot of interest and much more speculation. Severe technical difficulties are involved in attempting to do this. The largest mammal eggs such as that of the rabbit are 100 μm in diameter, or 1000 times smaller in volume than that of a frog's egg, and eggs of other species, such as a mouse or human, are about one-third of the volume of a rabbit's egg. In spite of this it is possible to inject material into mouse eggs so that some of them survive, using a pipette with a diameter of about 2μm (Lin 1966 and 1967; Wilson *et al.* 1972). Of mammalian cells still capable of division, the smallest have a nucleus with a diameter of about 5μm. Recently Bromhall (unpublished work) has succeeded in injecting rabbit eggs with a pipette of 6μm in internal diameter, but nuclei introduced into unfertilized eggs by this means have not succeeded in undergoing a normal first mitosis.

In view of the small size of mammal eggs, attempts have been made to introduce nuclei into them by cell fusion, which can be promoted by inactivated Sendai virus. In the first experiments of Graham (1969), spleen and bone marrow nuclei underwent some enlargement after fusion into fertilized or unfertilized mouse eggs, but frequently became pycnotic as the egg's chromosomes underwent their first mitosis. Some of the eggs with fused-in bone-marrow cells were grown to the 12-cell stage, but did not contain a chromosome marker of the donor nuclei.

The overall result of these experiments is that fused-in nuclei could be shown to persist in development up to, but not beyond, the two-cell stage. Subsequently Graham (1971) used improvements in technique to show that the failure of transplanted nuclei could not be accounted for by harmful effects of the fusion process. Graham suggests that mouse egg cytoplasm may fail to induce nuclear swelling, and the activation of DNA synthesis, in spleen and bone marrow nuclei fast enough for their chromosomes to divide normally at mitosis, a problem already encountered before in adult cell nuclei transplanted to frogs' eggs. Although there are no obvious reasons, in principle, why nuclear transplantation in mammals should not be achieved, and no doubt will be achieved in time, the technical difficulties involved have turned out to be even greater than were expected. In view of the likelihood that the factors which limit nuclear transfer success in mammals may be the same as those that apply to experiments with amphibia, it seems clear that frogs (in which up to 1000 nuclear transfers can be achieved in a day) should be used to develop further the technique of nuclear transplantation; the results can then be tested on mammals in which less than one-tenth as many transfers can be achieved within the same time and with the same effort.

B. The genetic analysis of animal somatic cells

Several vertebrate cell-lines have been isolated which differ from normal cells or from their parent stock in morphology, pattern of growth, nutritional requirements, resistance to drugs, etc. If the variant condition is stable when propagated through mitosis, these strains are often regarded as 'mutants', though the evidence that the variation is caused by a gene mutation is usually indirect (review by Krooth *et al.* 1968). Indeed an indication that mitotically-stable variations in cultured cells may not be genetic has come from experiments of Mezger-Freed (1971) involving the isolation of puromycin-resistant strains of a haploid frog cell line. Gene mutations should be detected much more often in a haploid cell-line (where the effect of a recessive gene is at once apparent) than in a diploid line (where a recessive gene's effect will be concealed by its dominant allele); mutations should also be obtained more often in mutagen-treated cells than in untreated cells. In fact the frequency with which puromycin-resistant variants were obtained was independent of mutagen treatment; such variants were obtained as often in haploid cells with one dose of a gene as in heteroploid cells with two or more doses of each gene per cell. These results imply that puromycin-resistance was not due to a gene mutation. Mezger-Freed (1971) suggests that puromycin-resistance may have been due to self-determining membrane units which govern the permeability of the membrane to puromycin.

The availability of a technique for transplanting nuclei from cultured cells opens up the possibility of carrying out a formal genetic analysis on a single somatic cell. The following diagram (Table 11) which is based on suggestions of Fischberg *et al.* (1963), shows how this could be done. In this scheme the use of germ-cell grafts would be unnecessary, if nuclear transfers from cultured cells commonly gave rise to fertile adults; in fact this

TABLE 11

The crossing of a single somatic cell with itself†

This diagram illustrates a sequence of procedures which could be used to carry out a formal mendelian analysis on a single cultured cell of *Xenopus laevis*.

Experimental procedure	*Reference to methods involved*
Single cultured cell	
↓	
Nuclear transfer to enucleated egg	
↓	
Blastula first-transfer embryo	Gurdon and Laskey
↓	(1970)
Serial nuclear transfers	
↓	
Clone of nuclear-transplant embryos	
↓	
Germ-cell grafts from nuclear-transplant larvae to wild-type larvae which were grown from fertilized eggs; wild-type (host) larvae include genetic males and females	
↓	Blackler and Fischberg
Adult male and female frogs containing sperm and eggs which are mitotic products of the original cultured cell nucleus	(1961)
↓	
Male × female mating gives F_1 progeny of original cultured cell	
↓	
F_1 tadpoles (genetically all males or all females) are sex-reversed by oestradiol treatment or by testis implantation	Mikamo and Witschi (1963); Gallien (1956)
↓	
F_1 male and female frogs are mated to give F_2 progeny of original cultured cell	

† A genetic nuclear marker (Fig. 6) could be used to provide proof that the progeny animals finally obtained are genetic descendants of the original cultured cell nucleus. This could be most simply achieved by using a 1-*nu* cultured cell as nuclear donor, and 2-*nu* wild-type material as recipient eggs for nuclear transfers and as host embryos for germ-cell grafts. Thus failure to remove all host embryo germ-cells (genetically 2-*nu*) would not affect the results.

happens only rarely by presently-available techniques. Germ-cell grafts help to alleviate this difficulty by removing the need for the *somatic* part of a nuclear-transplant embryo to be entirely normal. In addition germ-cell grafts save a generation by yielding both males and females in the F_1 progeny of the original cultured cell. It is very fortunate that the differentiation of germ-cells into male or female gametes is determined by the chromosomal constitution of the *somatic* gonad cells and not by the chromosome constitution of the germ-cells themselves (Humphrey 1933; Blackler 1965).

This scheme in Table 11 is rather complicated, and involves several steps which are technically difficult. No serious attempt has been made to carry out all of the above steps in one experimental series, though some parts of the scheme have been used for genetic analysis by Fischberg *et al.* (1963). Table 11 does however serve to illustrate the wide range of manipulative techniques that can be applied to amphibian embryos, and shows what is at present the only means by which a single somatic cell of an animal can be subjected to a formal genetic analysis. Fig. 7a shows a cultured cell 'transplant-frog'.

C. Useful technical information for microinjection experiments with eggs and oocytes of 'Xenopus'

C.1. *The construction, calibration, and handling of micropipettes*

The most important feature of these techniques is that they provide, in the author's experience, the simplest means of injecting nuclei or known volumes of fluid into living oocytes and eggs, in such a way as to have a minimally harmful effect on their survival or development. Using these methods it is possible to inject up to 1000 eggs or oocytes in a day. This would not be possible with more refined and expensive equipment, which could however be used to inject more sensitive eggs of other species, or to deliver more accurately calibrated volumes.

Figure 28, a–c, illustrates a procedure for constructing micropipettes. Pipettes for fluid injection and nuclear transplantation require to be made differently. For fluid injections, the intermediate section (diameter *ca.* 300 μm) which carries the

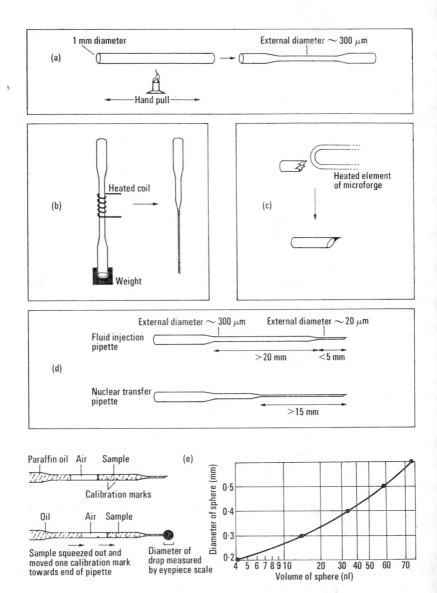

FIG. 28 The construction and calibration of microinjection pipettes. (a), (b) First and second pulls to draw out glass tubing, (c) shaping of the pipette tip by a microforge, (d) relative lengths of the fine and intermediate sections of pipettes for fluid injection and nuclear transplantation; (e) rapid method for calibrating fluid-injection pipettes. The pipette shown is calibrated so that four oocytes or eggs can be injected without refilling the pipette.

calibration marks, must be parallel-sided for at least 20 mm, and the fine section must be short so that the calibrated section remains in the field of view when the pipette tip is inside an egg (Fig. 28d). For nuclear transplantation, the fine (tip) section must be long to give good control over the movement of the nucleus, and the intermediate section need not be in the field of view. Pipettes can be rapidly calibrated to deliver volumes ranging from 10 to 100 nl, with an accuracy of $\pm$ 10 per cent, using the procedure shown in Fig. 28 e. An Agla syringe (Wellcome Reagents Ltd., Beckenham, Kent, England) should be connected to the micropipette by means of flexible polythene tubing of 1 mm diameter. The whole system should be filled with liquid paraffin (of specific gravity 0·85 g/ml) up to the point shown in Fig. 28e. The micropipette is best held, and its movement controlled, by a simple manipulator (c.g. those of Singer Instruments, London Road, Reading; or of Micro-instruments Ltd., Little Clarendon Street, Oxford, England). Injection is most conveniently carried out under a stereo-microscope at a magnification of about 15 × for fluid injections, but at a greater magnification for nuclear transfers.

C.2. *A medium for culture and injection*

Table 12 gives the constitution of the medium which has proved very satisfactory for culturing oocytes, as well as for nuclear transplantation. Injected eggs from which jelly has

TABLE 12

Incubation medium for injected oocytes and eggs of Xenopus†

Substance	Concentration (mM)	Grams per litre
NaCl	88	5·13
KCl	1·0	0·075
NaHCO$_3$	2·4	0·20
MgSO$_4$·7H$_2$O	0·82	0·20
Ca(NO$_3$)$_2$·4H$_2$O	0·33	0·08
CaCl$_2$·6H$_2$O	0·41	0·09
Benzyl penicillin	—	0·01
Streptomycin sulphate	—	0·01
Tris, HCl— pH 7·6	7·5	0·91

† This medium has been progressively modified from that of Barth and Barth (1959), from which it now differs in several respects.

been removed (partially by U.V. irradiation or completely by a 4-min. immersion in culture medium containing 2 per cent cysteine hydrochloride and brought to pH 8·1 with NaOH) must be transferred during cleavage to normal culture medium ten times diluted with water, since the full strength solution causes abnormal gastrulation. For injecting material, such as mRNA, into oocytes, a suitable medium is 88 mM NaCl, 15 mM Tris–HCl, pH 7·6. However the composition of the injection medium is not critical when an oocyte or egg is injected with less than 35 nl.

C.3. *The preparation, injection, and labelling of oocytes and eggs*

To obtain oocytes, a piece of ovary is removed from a killed or anaesthetized female, and at once separated into small fragments each containing not more than 30 full-sized oocytes. These fragments are spaced out, so that they are not contiguous, in a shallow dish of culture medium, and if maintained in this condition at 18°C these oocytes may be used for injection over the next few days. For injection, fragments of ovary are dissected into small clusters containing up to 5 full-sized oocytes. The cluster is placed without medium on a dry microscope slide, each oocyte is injected, and the cluster returned to the incubation medium. It is usually necessary to steady each oocyte with forceps as it is injected.

To procure eggs, a female is injected subcutaneously with mammalian gonadotropic hormone, obtainable commercially. The best results are obtained with a priming dose of about 250 I.U. given 16 hours before the time of intended laying, and a final dose of about 400 I.U. given 8 hours before this time. The optimal temperature for laying is 23°C. For further information concerning the handling of animals and the production of fertilized eggs, see Gurdon (1967 b). Eggs of *Xenopus* are covered with a sticky impenetrable jelly. This is partially removed, and penetration greatly facilitated, by exposure to u.v. radiation. The dose required varies from one batch of eggs to another, but is usually in the range of 150 000 ergs cm^2 given over 10 secs. For injection, eggs are placed, as dry as possible, on a microscope slide, exposed to U.V. radiation, and injected. The jelly sticks to the glass slide and makes it unnecessary to hold the eggs with forceps.

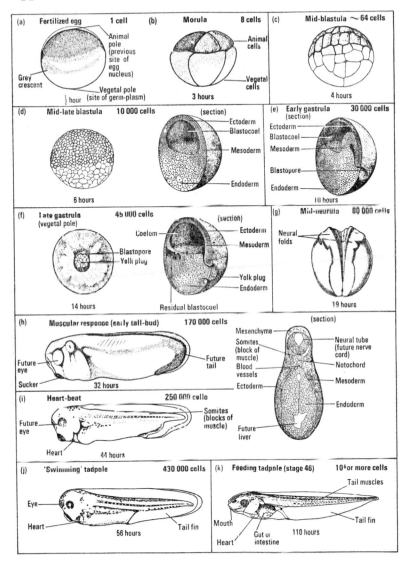

FIG. 29 Normal development of *Xenopus laevis*, showing morphological stages, cell number, and principal regions of an embryo. The number of hours shown at each stage is the time between fertilization and the stage shown if embryos are incubated at 23°C. (Whole embryo drawings are from Nieuwkoop and Faber 1956, *q.v.* for further details.)

It is important to take both oocytes and eggs out of culture medium for injection, since this reduces leakage, and helps to keep the cells in position during injection. It is however essential to prevent eggs or oocytes from drying during injection, and this will usually take place after about 2 minutes. Therefore only as many cells should be placed on a slide at one time as can be injected before their surface becomes dry.

The penetration of labelled amino acids or nucleosides into *eggs* or *embryos* is very largely inhibited by their jelly coat, and these precursors must be injected. Methods of labelling with $^{32}PO_4$ or $^{14}CO_2$ are referred to by Gurdon (1967b). *Oocytes* absorb labelled amino acids from their incubation medium very rapidly, and labelled nucleosides quite satisfactorily. Only if the desired labelling period is less than one hour for amino acids or less than 10 hours for nucleosides is it worth injecting the label.

C.4. *The normal development of* Xenopus laevis

Since reference has frequently been made in this book to stages of development and to regions of an embryo, the principal events in early amphibian development are shown in Fig. 29. The early development of most amphibians and many other animal species resembles that shown in this figure in broad outline. The early development of mammals is morphologically very different, and this appears to be related to the need to provide large amounts of yolk for nutrition in the early development of non-mammalian embryos.

Glossary

The explanations below are appropriate to the use of words in this book; in other contexts, a word may have a wider meaning. The glossary does not include chemical procedures or names of chemicals and animals. Regions of an embryo are illustrated in Fig. 29.

Actinomycin: a drug which inhibits RNA synthesis (transcription).

Activation (of an egg): a set of responses made by an unfertilized egg to fertilization or penetration by a pipette; these include the rupture of cortical granules, the lifting off of a vitelline membrane and free rotation of the egg by the influence of gravity.

Adenocarcinoma (renal): a pathologist's term for cancer of the kidney epithelium; it is usually associated, in frogs, with the presence of virus particles belonging to the Herpes group.

Allele: one of two or more alternative forms of a gene.

Animal pole: the end of an egg's surface with the least concentration of yolk; the animal pole of an amphibian egg usually contains the egg nucleus. Cells formed from the animal pole become ectoderm.

Antenna: appendage attached to the head of insects.

Antibody: a protein produced by an animal's plasma cells in response to foreign molecules to which it binds tightly and whose activity it neutralizes.

Anucleolate: lacking nucleoli—see Fig. 6.

Ascites cells: cancer cells, which grow freely in an abnormal accumulation of fluid in the body cavity of animals such as mice.

Autoimmune disease: one in which humans make antibodies against their own cell components; these antibodies may turn out to have the valuable property of reacting specifically with certain proteins of the nucleolus or nuclear sap.

Autoradiography: a technique by which a radiosensitive film is spread over fixed material; development of the film shows the exact distribution of labelled molecules or components within cells.

Bacteriophage: a virus which grows in bacteria.

Balbiani ring: a localized expansion of a polytene chromosome, believed to reflect transcriptional activity of genes in that region.

Blastocyst: an early (pre-implantation) stage of mammalian development; a hollow ball of 100–1000 cells (according to species), some of which will form the embryo.

Blastoderm stage: an early developmental stage; in insects, about 4000 cells are arranged in a single layer just inside the surface of the egg, the interior being filled with yolky cytoplasm. See Fig. 20.

Blastomere: a single large cell of a very early embryo, e.g. one cell of an eight-cell embryo.

Buoyant density (gradient): used in methods which distinguish DNA molecules according to the position in a density gradient where they come to rest independently of the length of time for which centrifugation is continued.

Callus: a disorganized mass of undifferentiated (parenchyma) plant cells, which may accumulate at a wounded surface, or grow out from pieces of plant tissue placed in culture.

Centromere: the region of a chromosome which is attached to the spindle at mitosis, and which is usually seen as a constriction in mitotic chromosomes.

Chromatin: the fibrous nucleo-protein complex of chromosomes; when isolated, it can be used to study RNA synthesis *in vitro*.

Chromomere: one of several dark-staining granules seen on meiotic chromosomes.

Clone: a group of genetically-identical individuals.

Coelomic mesoderm: the mesoderm which lines the body cavity of an embryo.

Colchicine: a substance which inhibits spindle formation and so arrests cells in mitosis when they possess condensed (and easily counted) chromosomes.

Collagen: A protein secreted as long fibres which are cross-linked together outside the cell; these constitute a major component of the material which exists between cells in many different tissues; probably synthesized by fibroblasts.

Cornea: a transparent cellular layer, about 0·5 mm thick, which covers the exposed surface of a vertebrate eye.

Cortex (of eggs): a thin layer of cytoplasm in the outer surface of eggs and early embryos.

Cotyledon: the first leaf (or first two leaves) of a developing plant, already present in the seed, and usually differing in shape from later leaves.

Crystallin: the major protein component of vertebrate lens cells; it consists of several closely related polypeptides.

Cycloheximide: a drug which inhibits protein synthesis in animal cells but not in bacteria, by a mechanism which is different from that of puromycin but which is not well understood.

Determined: an embryological term, meaning committed to differentiate in one of a limited number of ways.

Dialysable: able to pass through a dialysis membrane which is not penetrable by molecules larger than about 10 000 in molecular weight.

Diploid: containing two complete sets of chromosomes per nucleus, as do most plant and animal cells.

Disjunction: the separation of daughter chromosomes towards the end of mitosis.

Ecdysis, ecdysone: the moulting process, by which an insect larva changes into a larger larva or a pupa; the process is stimulated by the hormone 'ecdysone'.

Ectoderm: the layer of cells in an embryo which develop mainly into parts of the skin and the nervous system, and which is derived from the least yolky (animal) region of an egg.

Elongation: the process by which amino acids are added to a growing protein chain during protein synthesis.

Embryogenesis: the process by which an embryo is formed; i.e. early development.

Endoderm: the layer of cells in an embryo which gives rise to parts of the intestine, lungs, liver, etc., and which is derived from the most yolky (vegetal) part of an embryo.

Endogenous: of internal origin, e.g. messenger RNA molecules synthesized in an egg as opposed to those introduced by injection.

Enucleate: lacking a nucleus.

Epidermis (of plants): the outer layer of cells which surrounds plant tissues; leaf epidermis includes openings called stomata.

Epithelium: a covering sheet of cells, such as those which line the intestine.

Erythroblast: an incompletely differentiated cell which undergoes mitosis to give rise to mature red blood cells; in vertebrates, erythroblasts are typically located in bone marrow.

Erythrocyte: a mature red blood cell.

Explant: a piece of tissue which is cut out from a plant or animal and grown in culture.

Fibroblasts: cells of elongated or branching shape found in vertebrate connective tissue; they are present as a minority cell-type in most adult tissues, and apparently synthesize collagen.

Fibroin: a major protein of silk; it constitutes most of the protein synthesized by the posterior silk-gland cells of certain moth larvae.

Follicle (cells): small cells which surround an oocyte; in amphibia, a full-sized oocyte is covered by many hundred such cells, but each is less in volume than 10^{-4} of a mature oocyte.

Gamete: a reproductive cell, i.e. a sperm or egg cell.

Ganglion: a small mass of nervous tissue containing numerous cell bodies.

Genome: a collective term for all genes contained in a single (haploid) set of chromosomes.

Germ-cells, Germ-line, Germ-plasm: germ-cells are the embryonic cells from which gametes are formed, the germ-cells and gametes together constituting the germ-line. Germ-cells are derived from those early blastula cells which acquired a fraction of the special kind of cytoplasm (germ-plasm or pole plasm) located near the vegetal pole of an unfertilized egg.

Germinal vesicle: the large nucleus of an oocyte, containing lampbrush chromosomes.

Globin: a protein of 16 000 MW; together with haem, it constitutes haemoglobin, and amounts to much more than 90 per cent of the protein synthesized by red blood cells.

Gonadotropic hormones: a group of protein hormones, usually secreted by the pituitary gland; they stimulate growth and maturation of germ-cells into eggs and sperm, and control gonad activity.

Grey crescent: a region of cytoplasm which appears in amphibian eggs soon after fertilization. This is the site of the future blastopore (Fig. 29) and is usually formed opposite the point at which a sperm enters an egg.

Haemin: a porphyrin (non-protein) molecule which combines with globin to form haemoglobin.

Halteres: small appendages attached to the thorax of flies in place of the hind wings.

Haploid: containing one complete set of chromosomes per nucleus.

Hepatoma: a cancer of the liver.

Heterochromatin: regions of non-mitotic chromosomes which stain more densely than others and which are usually inactive in RNA synthesis.

Heterozygous: see homozygous.

Histone: a class of basic proteins characteristically associated with nuclear DNA and present in a similar amount per nucleus. There are five main classes of histones which range from 10 000–20 000 in molecular weight.

Homozygous: a diploid organism with the same allele at the equivalent place in both homologous chromosomes is homozygous for this allele. If it has a mutant allele on one chromosome but the wild-type allele on the other homologous chromosome, it is heterozygous.

Hybridization (of molecules): explained on page 9.

Imaginal disc: a group of cells in insect larvae which are destined to differentiate, in the adult, into a particular structure such as a wing, leg, or antenna etc.

Immunoglobulin: a protein having the structural properties, but not necessarily the activity, of an antibody.

Initiation factors: molecules which are needed for binding ribosomes to messenger RNA, and for initiating the translation of messages. There are several kinds of initiation factors, several of which can be extracted from polysomes with 1·0 M KCl.

Inversion: a change in the position of part of a chromosome, so that genes in this region are arranged in inverse order compared to normal.

Iris: a sheet of pigmented cells forming a diaphragm in front of the lens of vertebrate eyes.

Isoelectric point (of a protein): the pH at which a protein does not migrate in an electric field; proteins often differ in this respect.

Karyotype: the number and morphological characteristics of the chromosomes of a cell.

Keratin: the fibrous protein of skin cells, composed in most vertebrates of several related polypeptides.

Lampbrush (chromosomes): the highly extended chromosomes of animal oocytes, notable on account of their morphological details visible under a light microscope.

Leukaemia: cancer of the white blood cells.

Lysate: the cytoplasmic contents of ruptured cells; a reticulocyte lysate may be used to translate mRNA *in vitro*.

Malignant: having the uncontrolled growth characteristics of cancer cells.

Maturation (of oocytes): the process by which a fully grown oocyte is induced by hormone to complete meiosis and become a fertilizable egg.

Meiosis: the process by which the chromosome number of an oocyte or spermatocyte is reduced to the haploid amount; amphibian oocytes are in meiosis during their growth, until stimulated by hormones to undergo maturation.

Melanophore: a cell containing the pigment melanin.

Mesoderm: a layer of cells often derived from the equatorial region of an egg; it moves inside an embryo at gastrulation and forms muscle, blood, and parts of many adult organs.

Mesenchyme: embryonic connective tissue, which is derived from mesoderm, and which contains branching cells in gelatinous matrix.

Message: = messenger RNA or mRNA.

Metamorphosis (of insects and frogs) a stage in life when a larva or tadpole changes into an adult form, an event in which hormone action is commonly involved.

Microsomes: ribosomes attached to intracellular membranes; microsomes can be collected as a pellet by centrifuging homogenized cells at a moderate speed.

Microtubule: structures which are involved in intracellular movement and which characteristically constitute the spindle of mitotic cells.

Mitosis: the process by which a cell divides into two daughter cells with the same chromosome number as the parent cell.

Monosome: a single ribosome with mRNA attached to it.

Mosaics (nuclear): an organism consisting of cells with different kinds of chromosome constitution (some normal, some abnormal).

Mutagen: an agent, such as irradiation or certain chemicals, which induces mutation.

Myeloma: a cancer of antibody-secreting cells which synthesize mainly one kind of immunoglobulin.

Myoblast: an apparently undifferentiated cell-type which does not synthesize myosin to a detectable extent; myoblasts are 'determined' to differentiate into multinucleate muscle fibres after several such cells have fused together.

Myosin: a large protein involved in muscular contraction and synthesized in large amounts by muscle cells.

Nascent (protein chains): proteins in the process of being formed on a messenger RNA template in polysomes.

Neuroblasts: an apparently undifferentiated cell of the embryonic brain which will become a nerve-cell. In many animals, a neuroblast undergoes mitosis, one of the daughter cells differentiating into a nerve cell, and the other remaining a neuroblast which can again divide in the same way.

Non-malignant: lacking the uncontrolled growth characteristics of cancer cells.

Notochord: a sheathed rod of vacuolated cells located just ventral to the nerve cord of a vertebrate embryo, and involved in the formation of backbone.

Nuclear marker: a characteristic by which nuclei of one genetic type can be distinguished from those of other types—see Fig. 6.

Nuclear mosaic: see under mosaic.

Nuclear-transplant embryo: an embryo grown from a transplanted nucleus injected into an unfertilized enucleated egg.

Nucleolus: a microscopically visible object inside a nucleus, formed on the region of a chromosome at which ribosomal RNA genes are located; involved in ribosome synthesis.

Oocyte: a growing germ-cell which is located in the ovary and is converted into an egg at maturation. It has the genetic constitution of the mother, and for much of its growth has lampbrush chromosomes. See Fig. 3.

Oogenesis: the growth in the ovary of a germ-cell into the large yolky oocyte; this is then converted during maturation into an egg (see Fig. 3).

Orcein: a stain for mitotic chromosomes, used to determine a karyotype.

Ovalbumin: a protein synthesized in large amounts by hen oviduct cells; it is a major component of hens' egg white.

Ovary: the organ in which oocytes grow.

Ovulation: the first event in maturation when an oocyte is released from the ovary by the rupture of its follicle cells (see Fig. 3).

Parenchyma: a tissue containing unspecialized cells; these compose the pith and cortex of plants, and are capable of division.

Phenotypic (characteristics): the observable properties or a cell or organism as opposed to its inherent genetic properties.

Phloem: the principal food-conducting tissue of the vascular plants; phloem tissue consists of different cell-types which include parenchyma cells as well as sieve cells, fibre cells, and sclereids.

Pith: tissue which occupies the centre of plant stems inside the vascular tissue; it consists mainly of unspecialized parenchyma cells.

Pole plasm: a localized region of cytoplasm located at one end of an insect egg; cells in which the pole plasm becomes included during cleavage are thereby determined to form germ-cells.

Polar bodies: haploid cells extruded from an oocyte as it undergoes the first and second meiotic divisions during maturation into an egg.

Polar lobe: a lobe of cytoplasm which projects from one cell of early mollusc embryos; this cytoplasm is eventually included in, and is essential for the differentiation of, those cells which form the heart and intestine.

Polymerase (RNA, DNA): enzymes which promote the assembly of nucleotides or deoxynucleotides into RNA or DNA on a DNA template, processes known respectively as transcription and replication.

Polyoma: a small DNA-containing virus which normally infects mammalian cells, and which can cause cancer of infected tissues.

Polyploid: containing more than two complete sets of chromosomes per nucleus.

Polysome: a molecule of messenger RNA with a ribosome attached at each point along it where the message is being translated into a protein.

Polytene: many threaded; insect polytene chromosomes have about 1000 replicas of the chromosome arranged alongside each other in a bundle.

Promellitin: a protein which is the precursor of mellitin, a peptide synthesized by the venom glands of queen bees.

Pronuclei: the egg and sperm nuclei (both haploid) which are present in a newly fertilized egg; these later adhere to each other to form the zygote nucleus.

Protocollagen: a newly synthesized protein which is later modified by hydroxylation of some proline residues, and in some other ways, to form collagen.

Pulse-chase (labelling): a procedure in which cells are allowed to incorporate labelled molecules, and then placed in unlabelled medium; the stability of a molecule can be estimated from the rate at which label disappears from it during the chase.

Pupa: an immotile stage of development which exists, in most insects, between the larval and adult stages.

Puromycin: a drug which inhibits protein synthesis in all cells; it is incorporated, in place of a normal amino acid-tRNA, into a nascent protein chain, which is then released prematurely from the ribosome to which it was attached.

Recessive (gene): one whose effect is not observed when present in the same nucleus as a 'dominant' gene (allele) in the same position on the other set of chromosomes. A new mutation is almost always recessive to the wild-type gene.

Rescue (experiments): ones in which a genetic or experimentally induced deficiency of development is corrected by the injection of cytoplasm from normal eggs.

Reticulocyte: a red blood cell just before the final stage in its differentiation; reticulocytes (present in blood of anaemic animals) have lost their nuclei but are highly active in haemoglobin synthesis and are a favoured source of globin messenger RNA.

Retina: a multi-layered sheet of cells in the eye; it includes light-sensitive cells, pigment cells, and neurons, and is derived developmentally from an outgrowth of the brain to which it transmits the effects of light stimuli.

Ribosomal DNA: the part of a cells' DNA which codes for ribosomal RNA—see Table 1 and Fig. 2.

Semi-conservative (replication of DNA): occurs when each two-stranded daughter DNA molecule contains one complete strand of provide donor nuclei for further nuclear transfers (see Fig. 5).

Serial (nuclear-transfer experiment): an embryo which has itself resulted from a nuclear-transfer experiment is used to donor nuclei for further nuclear transfers (see Fig. 5).

Serum: the non-cellular component of blood, deprived of clotted constituents.

Somatic (cells): all cell-types in an individual other than those which are, or give rise to, eggs and sperm.

Spermatocyte: a type of cell contained in a testis; it undergoes a meiotic division to give haploid daughter sperm cells.

Spermine: a small positively-charged molecule (a polyamine), which is found in many animal tissues, and which has a stabilizing effect, *in vitro*, on cellular components including nucleic acids.

Spindle: the fibrous structure to which chromosomes are attached during mitosis; it includes microtubular proteins.

Stoma: an opening between epidermal cells of plant leaves; each stoma is surrounded by two guard cells.

Supercoiled (DNA): a configuration in which a double-stranded DNA helix is arranged in larger coils, and which is characteristic of complete molecules of polyoma virus DNA.

Template: as used here, DNA which is used for the transcription of RNA.

Termination: the incorporation of the last amino acid into a nascent protein chain and the release of the complete chain from the ribosome and mRNA.

Tetraploid: containing four complete chromosome sets per nucleus.

Transcription: the assembly of ribonucleotides into an RNA molecule which is coded for by one of the two strands of a DNA molecule.

Translation: the assembly of amino acids into a protein molecule in a sequence which is coded for by the order of nucleotides in a messenger RNA molecule.

Translocation: An event in which part of one chromosome is moved to another chromosome or to another part of the same chromosome; the relative position of genes is changed as a result.

Triploid: containing three complete chromosome sets per nucleus.

Turnover: the continuous synthesis and breakdown of molecules.

Vacuole: an intracellular fluid-filled space devoid of cytoplasmic particles and formed in most specialized plant cells.

Vegetal pole: the 'end' of an egg which contains the greatest concentration of yolk; cells at this end of the egg usually become endoderm and mesoderm.

Virion: the complete virus particle (as opposed to parts of the virus which can be extracted from infected cells).

Wild-type (gene): a type of gene which is found in wild populations of animals more commonly than any other variant of this gene (allele) able to occupy the equivalent site on homologous chromosomes.

X-chromosome: a particular chromosome which in most animals determines the sex of an individual according to whether it is present in one or two doses. In female mammals one of the two X-chromosomes present in diploid cells has all its genes inactive.

Zygote (nucleus): the fused or adjacent egg and sperm nuclei present in fertilized but still uncleaved eggs.

References

ABE, H. and YAMANA, K. (1970). The synthesis of 5-S RNA and its regulation during early embryogenesis of *Xenopus laevis*. *Biochim. Biophys. Acta*, **213**: 392–406. [78]

ARMS, K. (1968). Cytonucleoproteins in cleaving eggs of *Xenopus laevis*. *J. Embryol. exp. Morph.*, **20**: 367–74. [107]

AYDELOTTE, M. A. (1963). The effects of vitamin A and citral on epithelial differentiation *in vitro* 1. The chick tracheal epithelium. *J. Embryol. exp. Morph.*, **11**: 279–91. [19]

BACHVAROVA, R. and DAVIDSON, E. H. (1966). Nuclear activation at the onset of amphibian gastrulation. *J. exp. Zool.*, **163**: 285–96. [78]

BAGLIONI, C. (1963). Correlations between genetics and chemistry of human haemoglobins. In *Molecular Genetics*, vol. 1 (ed. J. H. Taylor), pp. 405–76. Academic Press, New York. [46]

BANTOCK, C. R. (1970). Experiments on chromosome elimination in the gall midge, *Mayetiola destructor*. *J. Embryol. exp. Morph.*, **24**: 257–86. [11, 87]

BARTH, L. G. and BARTH, L. J. (1959). Differentiation of cells of the *Rana pipiens* gastrula in unconditioned medium. *J. Embryol. exp. Morph.*, **7**: 210–22. [123]

BARRY, J. M. and MERRIAM, R. W. (1972). Swelling of hen erythrocyte nuclei in cytoplasm from *Xenopus laevis* eggs. *Exp. Cell Res.*, **71**: 90–6. [102]

BECK, J. S. (1962). The behaviour of certain nuclear antigens in mitosis. *Exp. Cell Res.*, **28**: 400–18. [115]

BEERMAN, W. (1956). Nuclear differentiation and functional morphology of chromosomes. *Cold Spring Harb. Symp.*, **21**: 217–30. [76]

—— (1963). Cytological aspects of information transfer in cellular differentiation. *Amer. Zool.*, **3**: 23–32. [76]

—— (1973). Chromomeres and genes. In *Results and problems in cell differentiation*, vol. 4, *Developmental studies of giant chromosomes*. Springer Verlag, Berlin. [76]

BEETSCHEN, J.-C. (1970). Existence d'un effet maternel dans la descendance des femelles de l'amphibien urodèle *Pleurodeles Waltlii*, homozygotes pour le facteur récessif *ac* (ascites caudale). *C. r. Acad. Sci. Paris*, **270**: 855–8. [91]

—— and JAYLET, A. (1965). Sur un facteur récessif semi-létale determinant l'apparition d'ascite caudale (*ac*), chez le Triton *Pleurodeles Waltlii*. *C. r. Acad. Sci. Paris*, **261**: 5675–8. [91]

BERENDES, H. D. (1969). Induction and control of puffing. *Ann. d'Embryol. Morph.*, **1**: 153–64. [76]

BERNS, A. J. M., STROUS, G. J. A. M., and BLOEMENDAL, H. (1972a). Heterologous *in vitro* synthesis of lens α-crystallin polypeptide. *Nature (Lond.) New Biol.*, **236**: 7–9. [50]

——, KRAAIKAMP, M., BLOEMENDAL, H., and LANE, C. D. (1972b). Calf crystallin synthesis in frog cells: the translation of lens-cell 14s RNA in oocytes. *Proc. natn. Acad. Sci., U.S.A.*, **69**: 1606–9. [58, 72]

BHARGAVA, P. M. and SHANMUGAM, G. (1971). Uptake of non-viral nucleic acids by mammalian cells. *Progr. Nucleic Acid Res.*, **11**: 103–92. [54]

BIRNSTIEL, M. L., CHIPCHASE, M., and SPEIRS, J. (1971). The ribosomal RNA cistrons. *Progr. Nucleic Acid Res.*, **11**: 351–89. [5, 6, 7, 85]

BLACKLER, A. W. (1962). Transfer of primordial germ-cells between two subspecies of *Xenopus laevis*. *J. Embryol. exp. Morph.*, **10**: 641–51. [87]

—— (1965). Germ-cell transfer and sex-ratio in *Xenopus laevis*. *J. Embryol. exp. Morph.*, **13**: 51–61. [121]

—— and FISCHBERG, M. (1961). Transfer of primordial germ-cells in *Xenopus laevis*. *J. Embryol. exp. Morph.*, **9**: 634–47. [120]

BOLUND, L., RINGERTZ, N. R., and HARRIS, H. (1969). Changes in the cytochemical properties of erythrocyte nuclei reactivated by cell fusion. *J. Cell Sci.*, **4**: 71–87. [107]

BOUNOURE, L., AUBRY, R., and HUCK, M.-L. (1954). Nouvelles recherches expérimentales sur les origines de la lignée reproductrice chez la Grenouille rousse. *J. Embryol. exp. Morph.*, **2**: 245–63. [87]

BOVERI, T. (1899). Die Entwicklung von *Ascaris megalocephala* mit besonderer Rücksicht auf die Kernverhältnisse. *Festschrift FC von Kupfer*, Jena. [93]

BRACHET, J. and DENIS, H. (1963). Effects of actinomycin D on morphogenesis. *Nature*, **198**: 205–6. [83]

—— and HUBERT, E. (1972). Studies on nucleocytoplasmic interactions during early amphibian development. I. Localised destruction of the egg cortex. *J. Embryol. exp. Morph.*, **27**: 121–45. [87]

——, FICQ, A., and TENCER, R. (1963). Amino acid incorporation into proteins of nucleate and anucleate fragments of sea urchin eggs: effect of parthenogenetic activation. *Exp. Cell Res.*, **32**: 168–70. [42]

——, HUEZ, G., and HUBERT, E. (1973). Microinjection of rabbit

References   139

hemoglobin messenger RNA into amphibian oocytes and embryos. *Proc. natn. Acad. Sci., U.S.A.*, **70**: 543–7. [70]

BRIGGS, R. (1969). Genetic control of early embryonic development in the Mexican Axolotl, *Ambystoma mexicanum. Ann. d'Embryol. Morphogen.*, **1**: 105–13. [91]

—— and KING, T. J. (1952). Transplantation of living nuclei from blastula cells into enucleated frogs' eggs. *Proc. natn. Acad. Sci., U.S.A.*, **38**: 455–63. [16]

—— —— (1957). Changes in the nuclei of differentiating endoderm cells as revealed by nuclear transplantation. *J. Morphol.*, **100**: 269–312. [25, 28, 33]

—— and JUSTUS, J. T. (1968). Partial characterization of the component from normal eggs which corrects the maternal effect of gene o in the Mexican Axolotl. *J. exp. Zool.*, **147**: 105–16. [91, 92]

KING, T. J. and DIBERARDINO, M. A. (1960). Development of nuclear-transplant embryos of known chromosome complement following parabiosis with normal embryos. In *Symposium on germ-cells and development*, pp. 441–77. Inst. Intern. d'Embryol. Fondaz. Baselli, Milano. [30]

——, SIGNORET, J., and HUMPHREY, R. R. (1964). Transplantation of nuclei of various cell types from neurulae of the Mexican Axolotl. *Devel. Biol.*, **10**: 233–46. [25, 28, 32]

BROWN, D. D. (1965). RNA synthesis during early development. In *Developmental and metabolic control mechanisms and neoplasia*, pp. 219–34. Williams and Wilkins, Baltimore. [78]

—— (1967). The genes for ribosomal RNA and their transcription during amphibian development. *Curr. Top. devel. Biol.*, **2**: 47–73. [7, 85]

—— and DAWID, I. (1968). Specific gene amplification in oocytes. *Science*, **160**: 272–80. [5, 10]

—— and GURDON, J. B. (1964). Absence of ribosomal RNA synthesis in the anucleolate mutant of *Xenopus laevis. Proc. natn. Acad. Sci., U.S.A.*, **51**: 139–46. [20, 105]

—— and LITTNA, E. (1964). RNA synthesis during the development of *Xenopus laevis*, the South African Clawed Toad. *J. mol. Biol.*, **8**: 669–87. [78]

—— and LITTNA, E. (1966). Synthesis and accumulation of DNA-like RNA during embryogenesis of *Xenopus laevis. J. mol. Biol.*, **20**: 81–94. [78]

—— and WEBER, C. S. (1968). Gene linkage by RNA–DNA hybridization. I. Unique DNA sequences homologous to 4s RNA, 5s RNA, and ribosomal RNA. *J. mol. Biol.*, **34**: 661–80. [6, 10, 114]

BROWN, S. W. (1966). Heterochromatin. *Science*, **151**: 417–25. [5]

BRUN, R. B. and KOBEL, H. R. (1972). Des grenouilles métamorphosées obtenues par transplantation nucléaire à partir du prosencéphale et de l'épiderme larvaire de *Xenopus laevis. Rev. Suisse de Zool.*, **79**: 961–5. [23]

CALLAN, H. G. (1963). The nature of lampbrush chromosomes. *Intern. Rev. Cytol.*, **15**: 1–34. [**77**]

—— and LLOYD, L. (1960). Lampbrush chromosomes of crested newts *Triturus cristatus* (Laurenti). *Phil. Trans. R. Soc.*, B, **243**: 135–219. [**77**]

CARLSON, J. G. (1952). Microdissection studies of the dividing neuroblast of the grasshopper, *Chortophaga viridifasciata* (De Geer). *Chromosoma*, **5**: 199–220. [**88**]

CECCARINI, C., MAGGIO, R., and BARBATA, G. (1967). Aminoacyl-sRNA synthetases as possible regulators of protein synthesis in the embryo in the sea urchin *Paracentrotus lividus*. *Proc. natn. Acad. Sci.*, *U.S.A.*, **58**: 2235–9. [**42**]

CHAPMAN, V. M., WHITTEN, W. K., and RUDDLE, F. H. (1971). Expression of paternal glucose phosphate isomerase-1 (GPI-1) in preimplantation stages of mouse embryos. *Devel. Biol.*, **26**: 153–8. [**83**]

CHEN, D., SARID, S., and KATCHALSKI, E. (1968). Studies on the nature of messenger RNA in germinating wheat embryos. *Proc. natn. Acad. Sci.*, *U.S.A.*, **60**: 902–9. [**45**]

CLAYTON, R. M. (1970). Problems of differentiation in the vertebrate lens. *Curr. Top. devel. Biol.*, **5**: 115–80. [**15**]

CLEVER, U. (1961). Genaktivitäten in der Riesenchromosomen von *Chironomus tentans* und ihre Beziehungen zur Entwicklung. I. Genaktivierung durch Ecdyson. *Chromosoma.*, **12**: 607–75. [**76**]

CLOWES, F. A. L. and JUNIPER, B. E. (1968). *Plant cells*. Blackwell, Oxford. [**89**]

COHEN, B. B. (1971). Cell-free protein synthesis in mixed systems with components from Ascites cells and reticulocytes. *Biochim. Biophys. Acta*, **247**: 133–40 [**49**]

COLLIER, J. R. (1966). Transcription of genetic information in the spiralian embryo. *Curr. Top. in devel. Biol.*, **1**: 39–59. [**80, 83**]

COMANDON, J. and DE FONBRUNE, P. (1939). Greffe nucléaire totale, simple ou multiple, chez une Amibe. *C. r. Soc. Biol.*, *Paris*, **130**: 744–8. [**16**]

COON, H. G. (1966). Clonal stability and phenotypic expression of chick cartilage cells *in vitro*. *Proc. natn. Acad. Sci.*, *U.S.A.*, **55**: 66–73. [**14**]

CRIPPA, M. (1970). Regulatory factor for the transcription of the ribosomal genes in amphibian oocytes. *Nature*, **226**: 1138–40. [**104**]

CROUSE, H. V. and KEYL, H.-G. (1968). Extra replications in the 'DNA-puffs' of *Sciara coprophila*. *Chromosoma*, **25**: 357–64. [**9**]

CURTIS, A. S. G. (1962). Morphogenetic interactions before gastrulation in the amphibian *Xenopus laevis*—the cortical field. *J. Embryol. exp. Morph.*, **10**: 410–22. [**87**]

DANIELLI, J. F., LORCH, I. J., ORD, M. J., and WILSON, E. G. (1955). Nucleus and cytoplasm in cellular inheritance. *Nature*, **176**: 1114–1115. [**16**]

References 141

DAVIDSON, E. H. (1968). *Gene activity in early development.* Academic Press, New York. [86, 109]

—— and HOUGH, B. R. (1971). Genetic information in oocyte RNA. *J. mol. Biol.*, 56: 491–506. [81]

DAWID, I., BROWN, D. D., and REEDER, R. H. (1970). Composition and structure of chromosomal and amplified ribosomal DNAs of *Xenopus laevis. J. mol. Biol.*, 51: 341–60. [6, 7]

DEÁK, I., SIDEBOTTOM, E., and HARRIS, H. (1972). Further experiments on the role of the nucleolus in the expression of structural genes. *J. Cell Sci.*, 11: 379–91. [98]

DENIS, H. (1966). Gene expression in amphibian development. I. Release of the genetic information in growing embryos. *J. mol. Biol.*, 22: 285–304. [81]

DENNY, P. C. and TYLER, A. (1964). Activation of protein biosynthesis in non-nucleate fragments of sea urchins. *Biochim. Biophys. Res. Commun.*, 14: 245–9. [42]

DE TERRA, N. (1969). Cytoplasmic control over nuclear events of cell reproduction. *Intern. Rev. Cytol.*, 25: 1–30. [16]

DI BERARDINO, M. A. and KING, T. J. (1966). Nuclear transplantation of differentiated male germ cells. *Am. Zool.*, 6: 510. [29]

—— —— (1967). Development and cellular differentiation of neural nuclear-transplants of known karyotype. *Devel. Biol.*, 15: 102–28. [33]

DOBOS, P., KERR, I. M., and MARTIN, E. M. (1971). Synthesis of capsid and non-capsid viral proteins in response to encephalomyocarditis virus RNA in animal cell-free systems. *J. Virol.*, 8: 491–9. [41]

DRACH, J. C. and LINGREL, J. B. (1966). Function of reticulocyte ribonucleic acid in the *Escherichia coli* cell-free system. *Biochim. Biophys. Acta*, 129: 178–85. [41]

DUPRAW, E. J. (1967). The honeybee embryo. In *Methods in developmental Biology* (ed. F. Wilt and N. K. Wessells), pp. 183–217. T. Y. Crowell Co. New York. [16]

ECKER, R. E. and SMITH, L. D. (1971). The nature and fate of *Rana pipiens* proteins synthesized during maturation and early cleavage. *Devel. Biol.*, 24: 559–76. [107]

EGUCHI, G. and OKADA, T. S. (1973). Differentiation of lens tissue from the progeny of chick retinal pigment cells cultured *in vitro*: a demonstration of a switch of cell-types in clonal cell culture. *Proc. natn. Acad. Sci., U.S.A.*, 70: 1495–9. [15]

ELSDALE, T. R., FISCHBERG, M., and SMITH, S. (1958). A mutation that reduces nucleolar number in *Xenopus laevis. Exp. Cell Res.*, 14: 642–3. [19, 20]

——, GURDON, J. B., and FISCHBERG, M. (1960). A description of the technique for nuclear transplantation in *Xenopus laevis. J. Embryol. exp. Morph.*, 8: 437–44. [19]

EMERSON, C. P. and HUMPHREYS, T. (1970). Regulation of DNA-like RNA and the apparent activation of ribosomal RNA synthesis in sea urchin embryos: quantitative measurements of newly synthesized RNA. *Devel. Biol.*, **23**: 86–94. [80]

EMERSON, C. P. and HUMPHREYS, T. (1971). Ribosomal RNA synthesis and the multiple, atypical nucleoli in cleaving embryos. *Science*, **171**: 898–901. [78, 80]

EPHRUSSI, B. (1972). *Hybridisation of somatic cells.* Oxford University Press, England. [96]

FISCHBERG, M., BLACKLER, A. W., UEHLINGER, V., REYNAUD, J., DROIN, A., and STOCK, J. (1963). Nucleo-cytoplasmic control of development. In *Genetics today*, pp. 187–98. Pergamon Press, Oxford. [23, 120, 121]

FORD, P. J. (1971). Non-coordinated accumulation and synthesis of 5s ribonucleic acid by ovaries of *Xenopus laevis*. *Nature*, **233**: 561–564. [78]

—— and SOUTHERN, E. M. (1973). Different sequences for 5s RNA in kidney cells and ovaries of *Xenopus laevis*. *Nature (Lond.) New Biol.*, **241**: 7–12. [78]

GALL, J. G. (1968). Differential synthesis of the genes for ribosomal RNA during amphibian oogenesis. *Proc. natn. Acad. Sci., U.S.A.*, **60**: 553–60. [5]

—— (1969). The genes for ribosomal RNA during oogenesis. *Genetics, suppl.*, **61**: 121–32. [6, 10]

—— and CALLAN, H. G. (1962). ³H-uridine incorporation in lampbrush chromosomes. *Proc. natn. Acad. Sci., U.S.A.*, **48**: 562–70. [77]

—— and PARDUE, M. L. (1969). Formation and detection of RNA–DNA hybrid molecules in cytological preparations. *Proc. natn. Acad. Sci., U.S.A.*, **63**: 378–83. [10]

GALLIEN, L. (1956). Inversion expérimentale du sexe chez au anoure inférior *Xenopus laevis* daudin. Analyse des conséquences génétiques. *Biol. Bull. France Belg.*, **90**: 163–73. [120]

——, PICHERAL, B., and LACROIX, J.-C. (1963). Modifications de l'assortiment chromosomique chez les larves hypomorphes du Triton *Pleurodeles Waltlii* Michah obtenues par transplantation de noyaux. *C. r. Acad. Sci., Paris*, **257**: 1721–3. [31]

GAREN, A. and GEHRING, W. (1972). Repair of the lethal developmental defect in *Deep Orange* embryos of *Drosophila* by injection of normal egg cytoplasm. *Proc. natn. Acad. Sci., U.S.A.*, **69**: 2982–2985. [91]

GEIDUSCHEK, E. P. and HASELKORN, R. (1969). Messenger RNA. *Ann. Rev. Biochem.*, **38**: 647–76. [68]

GEIGY, R. (1931). Action de l'ultraviolet sur le pôle germinal dans l'oeuf de *Drosophila melanogaster*. *Rev. Suisse de Zool.*, **38**: 187–288. [87]

GELEHRTER, T. D. and TOMKINS, G. N. (1969). Control of tyrosine aminotransferase synthesis in tissue culture by a factor in serum. *Proc. natn. Acad. Sci., U.S.A.*, **64**: 723–30. [50, 51]

GEYER-DUSZYŃSKA, I. (1959). Experimental research on chromosome diminution in *Cecidomiidae* (Diptera). *J. exp. Zool.*, **141**: 391–448. [**11, 87**]

—— (1967). Experiments on nuclear transplantation in *Drosophila melanogaster*. *Rev. Suisse de Zool.*, **74**: 614–15. [**27**]

GIGLIONI, B., GIANNI, A. M., COMI, P., OTTOLENGHI, S., and RUNGGER, D. (1973). Translational control of globin synthesis by haemin in *Xenopus* oocytes. *Nature, New Biology*, **246**: 99–102.

GOLDSTEIN, L. and PRESCOTT, D. M. (1967). Protein interactions between nucleus and cytoplasm. In *The control of nuclear activity* (ed. L. Goldstein), pp. 273–98. Prentice Hall, Inc., U.S.A. (New Jersey). [**16, 107**]

GRÄSSMANN, A. and GRÄSSMANN, M. (1971). Über die Bildung von Melanin in Muskelzellen nach der direkten Übertragung von RNA aus Harding–Passey Melanomzellen. *Hoppe-Seyler's Z. Physiol. Chem.*, **352**: 527–32. [**54**]

GRAHAM, C. F. (1969). The fusion of cells with one- and two-cell mouse embryos. *Wistar Inst. Symp.*, **9**: 19–33. [**118**]

—— (1971). Virus assisted fusion of embryonic cells. *Karolinska Symposia in Reproductive Endocrinology*, **3**: 154 65. [**118**]

——, ARMS, K., and GURDON, J. B. (1966). The induction of DNA synthesis by frog egg cytoplasm. *Devel. Biol.*, **14**: 349–81. [**32, 94, 102**]

GREEN, H., GOLDBERG, B., SCHWARTZ, M., and BROWN, D. D. (1968). The synthesis of collagen during the development of *Xenopus laevis*. *Devel. Biol.*, **18**: 391–400. [**73**]

GRIENINGER, G. E. and ZETSCHE, K. (1972), Die Aktivität von Phosphoglucoseisomerase und Phosphoglucomutase während der Morphogenese kernhaltiger und kernlosen *Acetabularien*. *Planta (Berl.)*, **104**: 329–51. [**40**]

GRIPPO, P. and LO SCAVO, A. (1972). DNA polymerase activity during maturation in *Xenopus laevis* oocytes. *Biochem. Biophys. Res. Commun.*, **48**: 280–5. [**102**]

GROSS, K. W., RUDERMAN, J., JACOBS-LORENA, M., BAGLIONI, C., and GROSS, P. R. (1973). Cell-free synthesis of histones directed by messenger RNA from sea-urchin embryos. *Nature (Lond.) New Biol.*, **241**: 272–4. [**43**]

GROSS, P. R. (1967). The control of protein synthesis in embryonic development and differentiation. *Curr. Top. in devel. Biol.*, **2**: 1–47. [**42, 44, 83**]

——, MACKLIN, L. I., and MOYER, W. A. (1964). Templates for the first proteins of embryonic development. *Proc. natn. Acad. Sci., U.S.A.*, **51**, 407–13. [**42**]

GROSSBACH, U. (1969). Chromosomen-Aktivität und biochemische Zelldifferenzierung in den Speicheldrüsen von *Camptochironomus*. *Chromosoma*, **28**: 136–87. [**76, 77**]

GURDON, J. B. (1960*a*). The effects of ultraviolet irradiation on the uncleaved eggs of *Xenopus laevis*. *Quart. J. Microscop. Sci.*, **101**: 299–312. [**19**]

GURDON, J. B. (1960b). The developmental capacity of nuclei taken from differentiating endoderm cells of *Xenopus laevis*. *J. Embryol. exp. Morph.*, 8: 505–26. [28, 29, 30]

—— (1961a). Multiple genetically identical frogs. *J. Hered.*, 53: 5–9. [116, 117]

—— (1961b). The transplantation of nuclei between two subspecies of *Xenopus laevis*. *Heredity*, 16: 305–15. [117]

—— (1962a). The developmental capacity of nuclei taken from intestinal epithelium cells of feeding tadpoles. *J. Embryol. exp. Morph.*, 10: 622–40. [18, 22]

—— (1962b). Adult frogs derived from the nuclei of single somatic cells. *Devel. Biol.*, 4: 256–73. [22]

—— (1964). The transplantation of living cell nuclei. *Adv. Morphogen.*, 4: 1–43. [16, 117]

—— (1967a). On the origin and persistence of a cytoplasmic state inducing nuclear DNA synthesis in frogs' eggs. *Proc. natn. Acad. Sci.*, *U.S.A.*, 58: 545–52. [102, 105]

—— (1967b). African clawed frogs. In *Methods in developmental Biology* (ed. F. H. Wilt and N. K. Wessells), pp. 75–84. Crowell Co., New York. [124, 126]

—— (1968a). Transplanted nuclei and cell differentiation. *Scient. Am.*, 219(6): 24–35. [17, 19, 23]

—— (1968b). Changes in somatic cell nuclei inserted into growing and maturing amphibian oocytes. *J. Embryol. exp. Morph.*, 20: 401–14. [54, 94]

—— (1968c). Nucleic acid synthesis in embryos and its bearing on cell differentiation. *Essays in Biochemistry*, 4: 26–68. [78]

—— (1970). Nuclear transplantation and the control of gene activity in animal development. *Proc. R. Soc.*, *B.*, 176: 303–14. [107, 108]

—— and BROWN, D. D. (1965). Cytoplasmic regulation of RNA synthesis and nucleolus formation in developing embryos of *Xenopus laevis*. *J. mol. Biol.*, 12: 27–35. [103]

—— and LASKEY, R. A. (1970). The transplantation of nuclei from single cultured cells into enucleate frogs' eggs. *J. Embryol. exp. Morph.*, 24: 227–48. [18, 21, 22, 24, 32, 117, 120]

—— and SPEIGHT, V. A. (1969). The appearance of cytoplasmic DNA polymerase activity during the maturation of amphibian oocytes into eggs. *Exp. Cell Res.*, 55: 253–6. [99]

—— and UEHLINGER, V. (1966). 'Fertile' intestine nuclei. *Nature*, 210: 1240–1. [23]

—— and WOODLAND, H. R. (1968). The cytoplasmic control of nuclear activity in animal development. *Biol. Rev.*, 43: 233–67. [90]

—— —— (1969). The influence of the cytoplasm on the nucleus during cell differentiation with special reference to RNA synthesis during amphibian cleavage. *Proc. R. Soc.*, *B.*, 173: 99–111. [78, 80, 95, 97]

—— —— (1970). On the long-term control of nuclear activity during cell differentiation. *Curr. Top. in devel. Biol.*, **5**: 39–70. [**105, 115**]

—— —— (1974). *Xenopus*. In *Survey of Genetics* (ed. R. C. King). Plenum Press, U.S.A. Vol. 3 in press. [**20**]

——, BIRNSTIEL, M. L., and SPEIGHT, V. A. (1969). The replication of purified DNA introduced into living egg cytoplasm. *Biochim. Biophys. Acta. N*, **174**: 614–28. [**99**]

——, LANE, C. D., WOODLAND, H. R., and MARBAIX, G. (1971). Use of frog eggs and oocytes for the study of messenger RNA and its translation in living cells. *Nature*, **233**: 177–82. [**54, 57, 59**]

——, LINGREL, J. B., and MARBAIX, G. (1973). Message stability in injected frog oocytes. *J. mol. Biol.*, **80**: 539–551. [**56, 60, 69**]

HADORN, E. (1968). Transdetermination in cells. *Scient. Am.*, **219**(5): 110–20. [**14, 26**]

HÄMMERLING, J. (1953). Nucleo-cytoplasmic relationships in the development of *Acetabularia*. *Intern. Rev. Cytol.*, **2**: 475–98. [**40**]

HARRIS, S. E. and FORREST, H. S. (1967). RNA and DNA synthesis in developing eggs of the milkweed bug, *Oncopeltus fasciatus* (Dallas). *Science*, **156**: 1613–15. [**80**]

HARRIS, H., SIDEBOTTOM, E., GRACE, D. M., and BRAMWELL, M. (1969). The expression of genetic information: a study with hybrid animal cells. *J. Cell Sci.*, **4**: 499–526. [**81, 97, 98**]

HENNEN, S. (1963). Chromosomal and embryological analyses of nuclear changes occurring in embryos derived from transfers of nuclei between *Rana pipiens* and *Rana sylvatica*. *Devel. Biol.*, **6**: 133–83. [**31**]

—— (1970). Influence of spermine and reduced temperature on the ability of transplanted nuclei to promote normal development in eggs of *Rana pipiens*. *Proc. natn. Acad. Sci., U.S.A.*, **66**: 630–7. [**25, 28**]

HEYWOOD, S. (1969). Synthesis of myosin on heterologous ribosomes. *Cold Spring Harb. Symp.*, **34**: 799–803. [**41, 47**]

HILDRETH, P. E. and LUCCHESI, J. C. (1967). Fertilisation in *Drosophila*. III. A revaluation of the role of polyspermy in development of the mutant deep orange. *Devel. Biol.*, **15**: 536–52. [**84**]

HILL, R. N. and McCONKEY, E. (1972). Coordination of ribosomal RNA synthesis in vertebrate cells. *J. Cell Physiol.*, **79**: 15–26. [**103**]

HILLMAN, N. and HILLMAN, R. (1967). Competent chick ectoderm: non-specific response to RNA. *Science*, **155**: 1563–5. [**53**]

HOLTZER, H. and ABBOTT, J. (1968). Oscillations of the chondrogenic phenotype *in vitro*. In *The stability of the differentiated state* (ed. H. Ursprung), pp. 1–16. Springer Verlag. [**14**]

HULTIN, T. (1961). Activation of ribosomes in sea urchin eggs in response to fertilisation. *Exp. Cell Res.*, **25**: 405–17. [**42**]

HUMPHREY, R. R. (1933). The development and sex-differentiation of the gonad in the wood frog (*Rana sylvatica*) following extirpation or orthotopic implantation of the intermediate segment and adjacent mesoderm. *J. exp. Zool.*, **65**: 243–9. [**121**]

HUMPHREYS, T. (1969). Efficiency of translation of messenger RNA before and after fertilisation in sea-urchins. *Devel. Biol.*, **20**: 435–458. [42]

—— (1971). Measurements of messenger RNA entering polysomes upon fertilisation of sea urchin eggs. *Devel. Biol.*, **26**: 201–8. [42, 43, 65]

HUNT, T., HUNTER, T., and MUNRO, A. (1968). Control of haemoglobin synthesis: distribution of ribosomes on the messenger RNA for α and β chains. *J. mol. Biol.*, **36**: 31–45. [46, 48]

—— —— —— (1969). Control of haemoglobin synthesis: rate of translation of the messenger RNA for the α and β chains. *J. mol. Biol.*, **43**: 123–33. [59]

ILAN, J. and ILAN, J. (1971). Stage-specific initiation factors for protein synthesis during insect development. *Devel. Biol.*, **25**: 280–92. [49]

ILLMENSEE, K. (1968). Transplantation of embryonic nuclei into unfertilised eggs of *Drosophila melanogaster*. *Nature*, **219**: 1268–9. [27]

—— (1972). Developmental potencies of nuclei from cleavage, pre-blastoderm and syncytial blastoderm transplanted into unfertilised eggs of *Drosophila melanogaster*. *Wilhelm Roux' Archiv.*, **170**: 267–98. [26, 27]

JELINEK, W., ADESNIK, M., SALDITT, M., SHEINESS, M., WALL, R., MOLLOY, G., PHILIPSON, L., and DARNELL, J. E. (1973). Further evidence on the nuclear origin and transfer to the cytoplasm of polyadenylic acid sequences in mammalian cell RNA. *J. mol. Biol.*, **75**: 515–32. [75]

JONES, K. W. (1970). Chromosomal and nuclear location of mouse satellite DNA in individual cells. *Nature*, **225**: 912–15. [10]

KAFATOS, F. (1972). mRNA stability and cellular differentiation. *Karolinska Symp.*, **5**: 319–41. [70]

KEDES, L. H. and GROSS, P. R. (1969). Synthesis and function of messenger RNA during early embryonic development. *J. mol. Biol.*, **42**: 559–75. [44, 80]

—— ——, COGNETTI, G., and HUNTER, A. L. (1969). Synthesis of nuclear and chromosomal proteins on light polyribosomes during cleavage in the sea urchin embryo. *J. mol. Biol.*, **45**: 337–51. [45]

KING, T. J. and BRIGGS, R. (1955). Changes in the nuclei of differentiating gastrula cells, as demonstrated by nuclear transplantation. *Proc. natn. Acad. Sci., U.S.A.*, **41**: 321–5. [27]

—— —— (1956). Serial transplantation of embryonic nuclei. *Cold Spring Harb. Symp.*, **21**: 271–90. [29]

—— and MCKINNELL, R. G. (1960). An attempt to determine the developmental potentialities of the cancer cell nucleus by means of transplantation. In *Cell physiology of neoplasia*, pp. 591–617. University of Texas Press, U.S.A. [35]

—— and DIBERARDINO, M. A. (1965). Transplantation of nuclei from the frog renal adenocarcinoma. I. Development of tumor nuclear-transplant embryos. *Ann. N.Y. Acad. Sci.*, **126**: 115–26. [**18, 31, 35**]

KNOWLAND, J. S. and MILLER, L. (1970). Reduction of ribosomal RNA synthesis and ribosomal RNA genes in a mutant of *Xenopus laevis* which organises only a partial nucleolus. *J. mol. Biol.*, **53**: 321–38. [**20, 106**]

—— and GRAHAM, C. F. (1972). RNA synthesis at the two-cell stage of mouse development. *J. Embryol. exp. Morph.*, **27**: 167–76. [**78**]

—— GURDON, J. B., and LASKEY, R. A. (1974). Injection of messenger RNA into living cells and its application to the study of gene action in *Xenopus laevis*. In *RNA in development* (ed. M. C. Niu), North-Holland Publishing Co., Amsterdam. In the press. [**58, 72**]

KOBEL, H. R., BRUN, R. B., and FISCHBERG, M. (1973). Nuclear transplantation with melanophores, ciliated epidermal cells, and the established cell-line A-8 in *Xenopus laevis*. *J. Embryol. exp. Morph.*, **29**: 539–47. [**24**]

KROOTH, R. S., DARLINGTON, G. A., and VELASQUEZ, A. A. (1968). The genetics of cultured mammalian cells. *Ann. Rev. Genetics*, **2**: 141–64. [**119**]

KUMAR, K. V. and FRIEDMAN, D. L. (1972). Initiation of DNA synthesis in a HeLa cell-free system. *Nature (Lond.) New Biol.*, **239**: 71 6. [**102**]

LANDESMAN, R. (1972). Ribosomal RNA synthesis in pre- and post-gastrula embryos of *Xenopus laevis*. *Cell Differentiation*, **1**: 209–13. [**80**]

—— and GROSS, P. R. (1968). Patterns of macromolecule synthesis during development of *Xenopus laevis*. 1. Incorporation of radio-active precursors into dissociated embryos. *Devel. Biol.*, **18**: 571–589. [**103**]

———— (1969). Patterns of macromolecule synthesis during development of *Xenopus laevis*. II. Identification of the 40s precursor to ribosomal RNA. *Devel. Biol.*, **19**: 244–60. [**78**]

LANE, C. D. and KNOWLAND, J. S. (1974). The injection of RNA into living cells: the use of frog oocytes for the assay of messenger RNA and the study of the control of gene expression. In *Biochemistry of development*, vol. 3 (ed. R. Weber). Academic Press. [**54, 57, 58, 72, 73**]

——, MARBAIX, G., and GURDON, J. B. (1971). Rabbit haemoglobin synthesis in frog cells: the translation of reticulocyte 9s RNA in frog oocytes. *J. mol. Biol.*, **61**: 73–91. [**54, 55, 56**]

——, GREGORY, C. M., and MOREL, C. (1973). Duck-haemoglobin synthesis in frog cells. *Eur. J. Biochem.*, **34**: 219–27. [**56**]

LASKEY, R. A. and GURDON, J. B. (1970). Genetic content of adult somatic cells tested by nuclear transplantation from cultured cells. *Nature*, **228**: 1332–4. [**18**]

LASKEY, R. A. and GURDON, J. B. (1973). The induction of polyoma DNA synthesis by injection into frog egg cytoplasm. *Eur. J. Biochem.*, **37**: 467–71. [**99, 100, 101**]

—— —— and CRAWFORD, L. V. (1972). Translation of encephalomyocarditis viral RNA in oocytes of *Xenopus laevis. Proc. natn. Acad. Sci., U.S.A.*, **69**: 3665–9. [**58, 72**]

——, GERHART, J., and KNOWLAND, J. S. (1973). Inhibition of ribosomal RNA synthesis in neurula cells by extracts from blastulae of *Xenopus laevis. Devel. Biol.*, **33**: 241–7. [**103, 104**]

LIN, T. P. (1966). Microinjection into mouse eggs. *Science*, **151**: 333–337. [**118**]

—— (1967). Micropipetting cytoplasm from the mouse egg. *Nature*, **216**: 162–3. [**118**]

LINGREL, J. B. (1974). The translation of messenger RNA in cell-free systems. *MTP International Review of Science*, **8**. Butterworths Ltd., London. In the press. [**50**]

—— and WOODLAND, H. R. (1974). Initiation does not limit the rate of haemoglobin synthesis in message injected *Xenopus* oocytes. In preparation. [**59, 66**]

LOCKARD, R. E. and LINGREL, J. B. (1969). The synthesis of mouse haemoglobin β-chains in a rabbit reticulocyte cell-free system programmed with mouse reticulocyte 9s RNA. *Biochem. Biophys. Res. Commun.*, **37**: 204–12. [**41**]

LOCKSHIN, R. A. (1966). Insect embryogenesis: macromolecular synthesis during early development. *Science*, **154**: 775–6. [**80**]

LODISH, H. F. (1971). Alpha and beta globin messenger RNA: different amounts and rates of initiation of translation. *J. biol. Chem.*, **246**: 7131–8. [**47, 48, 65**]

—— and JACOBSEN, M. (1972). Regulation of haemoglobin synthesis: equal rates of translation and termination of α- and β-globin chains. *J. biol. Chem.*, **247**: 3622–9. [**47, 48**]

LONDON, I. M., TAVILL, A. S., VANDERHOFF, G. A., HUNT, T., and GRAYZEL, A. I. (1967). Erythroid cell differentiation and the synthesis and assembly of haemoglobin. *Devel. Biol. Suppl.*, **1**: 227–53. [**46**]

MACGREGOR, H. C. (1972). The nucleolus and its genes in amphibian oogenesis. *Biol. Rev.*, **47**: 177–210. [**6, 7**]

MACLEAN, N. and JURD., R. D. (1971). The haemoglobins of healthy and anaemic *Xenopus laevis. J. Cell Sci.*, **9**: 509–28. [**55**]

McCLINTOCK, B. (1956). Controlling elements and the gene. *Cold Spring Harb. Symp.*, **21**: 197–216. [**5**]

McKINNELL, R. G., DEGGINS, B. A., and LABAT, D. D. (1969). Transplantation of pluripotential nuclei from triploid frog tumors. *Science*, **165**: 394–6. [**25, 35**]

MAEYER-GUIGNARD, J. DE, MAEYER, E. DE, and MONTAGNIER, L. (1972). Interferon messenger RNA: translation in heterologous cells. *Proc. natn. Acad. Sci., U.S.A.*, **69**: 1203–7. (**54**)

MAGGIO, R., VITTORELLI, M. L., CAFFARELLI-MORMINO, I., and MON-

ROY, A. (1968). Dissociation of ribosomes of unfertilised eggs and embryos of sea urchins. *J. mol. Biol.*, **31**: 621–6. [42]

MAIRY, M. and DENIS, H. (1972). Recherches biochimiques sur l'oogenèse. 2. Assemblage des ribosomes pendant le grand accroissement des oocytes de *Xenopus laevis*. *Eur. J. Biochem.*, **25**: 535–43. [78]

MARBAIX, G. and LANE, C. D. (1972). Rabbit haemoglobin synthesis in frog cells. II. Further characterization of the products of translation of reticulocyte 9s RNA. *J. mol. Biol.*, **67**: 517–24. [56]

—— and GURDON, J. B. (1972). The effect of reticulocyte ribosome 'factors' on the translation of haemoglobin messenger RNA in living frog oocytes. *Biochim. Biophys. Acta*, **281**: 86–92. [65]

MARCUS, A. and FEELEY, J. (1966). Ribosome activation and polysome formation *in vitro*: requirement for ATP. *Proc. natn. Acad. Sci.*, *U.S.A.*, **56**: 1770–7 [45, 46]

MARKS, P. A. and KOVACH, J. S. (1966). Development of mammalian erythroid cells. *Curr. Top. devel. Biol.*, **1**: 213–52. [47, 68]

MARRÉ, E. (1967). Ribosome and enzyme changes during maturation and germination of the castor bean seed. *Curr. Top. devel. Biol.*, **2**: 75–105. [45]

MATHEWS, M. B. and KORNER, A. (1971). Mammalian cell-free protein synthesis directed by viral RNA. *Eur. J. Biochem.*, **17**: 328–338. [41]

MENDEL, G. (1866). Versuche über Pflanzenhybriden. *Verhandl. Naturforsch. Ver. Brünn.*, **4**: 3–47. [105]

MERRIAM, R. W. (1969). Movement of cytoplasmic proteins into nuclei induced to enlarge and initiate DNA or RNA synthesis. *J. Cell Sci.*, **5**: 333–49. [107]

METAFORA, S., FELICETTI, L., and GAMBINO, R. (1971). The mechanism of protein synthesis activation after fertilisation of sea urchin eggs. *Proc. natn. Acad. Sci.*, *U.S.A.*, **68**: 600–4. [42]

—— TERADA, M., DOW, L. W., MARKS, P. A., and BANK, A. (1972). Increased efficiency of exogenous messenger RNA translation in a Krebs Ascites cell lysate. *Proc. natn. Acad. Sci.*, *U.S.A.*, **69**: 1299–1303. [49]

MEZGER-FREED, L. (1971). Puromycin resistance in haploid and heteroploid frog cells: gene or membrane determined? *J. Cell Biol.*, **51**: 742–51. [119]

MIKAMO, K. and WITSCHI, E. (1963). Functional sex-reversal in genetic females of *Xenopus laevis*, induced by implanted testes. *Genetics*, **48**: 1411–21. [120]

MILLER, L. and GURDON, J. B. (1970). Mutations affecting the size of the nucleolus in *Xenopus laevis*. *Nature*, **227**: 1108–10. [20, 106]

—— and KNOWLAND, J. S. (1972). The number and activity of ribosomal RNA genes in *Xenopus laevis* embryos carrying partial deletions in both nucleolar organisers. *Biochem. Genet.*, **6**: 65–73. [20]

MINTZ, B. (1964). Formation of genetically mosaic mouse embryos and early development of Lethal (t^{12}/t^{12})–normal mosaics. *J. exp. Zool.*, **157**: 267–71. [80, 83]

MOAR, V. A., GURDON, J. B., and LANE, C. D. (1971). Translational capacity of living frog eggs and oocytes, as judged by messenger RNA injection. *J. mol. Biol.*, **61**: 93–104. [61, 63]

MONROY, A. (1965). *Chemistry and physiology of fertilisation.* Holt, New York. [42]

MOORE, J. A. (1955). Abnormal combinations of nuclear and cytoplasmic systems in frogs and toads. *Adv. Genet.*, **7**: 139–82. [83]

MORRISON, M. R., GORSKI, J., and LINGREL, J. B. (1972). The separation of mouse reticulocyte 9s RNA into fractions of different biological activity by hybridisation to poly (U)-cellulose. *Biochem. Biophys. Res. Commun.*, **49**: 775–81. [59]

MUELLER, G. C. (1969). Biochemical events in the animal cell cycle. *Federation Proc., U.S.A.*, **28**: 1780–9. [102]

MUGGLETON-HARRIS, A. L. and PEZZELLA, K. (1972). The ability of the lens cell nucleus to promote complete embryonic development and its applications to ophthalmic gerontology. *Exp. Gerontology*, **7**: 427–31. [25]

NEMER, M. (1963). Old and new RNA in the embryogenesis of the purple sea urchin. *Proc. natn. Acad. Sci., U.S.A.*, **50**: 230–5. [80]

NEWELL, P. C. (1971). The development of the cellular slime mould *Dictyostelium discoideum*: a model system for the study of cellular differentiation. *Essays in Biochemistry*, **7**: 87–126. [41]

NEYFAKH, A. A. (1971). Steps of realisation of genetic information in early development. *Curr. Top. in devel. Biol.*, **6**: 45–78. [83]

NIEUWKOOP, P. D. and FABER, J. (1956). Normal Table of *Xenopus laevis* (Daudin). North-Holland Publishing Co., Amsterdam. [125]

NIKITINA, L. A. (1964). Transfers of nuclei from the ectoderm and neural rudiments of developing embryos of *Bufo bufo, Rana arvalis*, and *Rana temporaria* into enucleated eggs of the same species. *Dokl. Nauk. S.S.S.R.*, **156**: 1468–71. [25]

NIKLAS, R. B. (1959). An experimental and descriptive study of chromosome elimination in *Miastor* spec. *Chromosoma*, **10**: 301–36. [12]

NIU, M. C. (1974). (ed.) *RNA in Development*, North-Holland Publishing Co., Amsterdam. [54]

—— and DESHPANDE, A. K. (1973). The development of tubular heart in RNA-treated post-nodal pieces of chick blastoderm. *J. Embryol. exp. Morph.*, **29**: 485–501. [53]

——, CORDOVA, C. C., and NIU, L. C. (1961). RNA-induced changes in mammalian cells. *Proc. natn. Acad. Sci., U.S.A.*, **47**: 1689–1700. [53]

OHNO, S., STENIUS, C., CHRISTIAN, L. C., and HARRIS, C. (1968). Synchronous activation of both parental alleles at the 6-PGD locus of Japanese quail embryos. *Biochem. Genet.*, **2**: 197–204. [83]

PACKMAN, S., AVIV, H., ROSS, J., and LEDER, P. (1972). A comparison of globin genes in duck reticulocytes and liver cells. *Biochem. Biophys. Res. Commun.*, **49**: 813–19. [9]

PAUL, J. (1970). DNA masking in mammalian chromatin: a molecular mechanism for determination of cell-type. *Curr. Top. in devel. Biol.*, **5**: 317–52. [114]

PAVAN, C. and DA CUNHA, A. B. (1969). Chromosomal activities in *Rhynchosciara* and other *Sciaridae. Ann. Rev. Genet.*, **3**: 425–50. [9]

PENMAN, S., SCHERRER, K., BECKER, Y., and DARNELL, J. E. (1963). Polyribosomes in normal and poliovirus-infected HeLa cells and their relationship to messenger RNA. *Proc. natn. Acad. Sci.*, *U.S.A.*, **49**: 654–62. [75]

PETERSON, J. A. and WEISS, M. (1972). Expression of differentiated functions in hepatoma cell hybrids: induction of mouse albumin production in rat hepatoma-mouse fibroblast hybrids. *Proc. natn. Acad. Sci., U.S.A.*, **69**: 571–5. [96]

PICHERAL, B. (1962). Capacités des noyeaux de cellules endodermiques embryonnaires à organiser un germe viable chez l'Urodele, *Pleurodeles Waltlii* Michah. *C. r. Acad. Sci., Paris*, **255**: 2500 2511. [25]

POULSON, D. F. (1945). Chromosomal control of embryogenesis in *Drosophila. Am. Nat.*, **79**: 340–63. [84, 85]

PRICHARD, P. M., PICCIANO, D. J., LAYCOCK, D. G., and ANDERSON, W. F. (1971). Translation of exogenous messenger RNA for hemoglobin on reticulocyte and liver ribosomes. *Proc. natn. Acad. Sci., U.S.A.*, **68**: 2752–6. [41, 49]

RAFF, R. A., COLOT, H. V., SELVIG, S. E., and GROSS, P. R. (1972). Oogenetic origin of messenger RNA for embryonic synthesis of microtubule proteins. *Nature*, **235**: 211–14. [45]

REYER, R. W. (1962). Regeneration in the amphibian eye. In *Regeneration* (ed. D. Rudnick), pp. 211 65. Ronald Press, New York. [15]

RHOADS, R. E., McKNIGHT, G. S., and SCHIMKE, R. T. (1971). Synthesis of ovalbumin in a rabbit reticulocyte cell-free system programmed with hen oviduct RNA. *J. biol. Chem.*, **246**: 7407–10. [50]

RINGERTZ, N. R., CARLSSON, S.-A., EGE, T., and BOLUND, L. (1971). Detection of human and chick nuclear antigens in nuclei of chick erythrocytes during reactivation in heterokaryons with HeLa cells. *Proc. natn. Acad. Sci., U.S.A.*, **68**: 3228–32. [115]

RITOSSA, R. F., ATWOOD, K. C., and SPEIGELMAN, S. (1966). A molecular explanation of the bobbed mutants of *Drosophila* as partial deficiencies of 'ribosomal' DNA. *Genetics*, **54**: 819–34. [106]

ROEDER, R. G. and RUTTER, W. J. (1970). Specific nucleolar and nucleoplasmic RNA polymerase. *Proc. natn. Acad. Sci., U.S.A.*, **65**: 675–82. [85]

——, REEDER, R. H., and BROWN, D. D. (1972). Multiple forms of RNA polymerase in *Xenopus laevis*: their relationship to RNA synthesis *in vivo* and their fidelity of transcription *in vitro*. *Cold Spring Harb. Symp.*, **35**: 727–35. [**85**]

ROLLINS, J. W. and FLICKINGER, R. A. (1972). Collagen synthesis in *Xenopus* oocytes after injection of nuclear or polysomal RNA of frog embryos. *Science*, **178**: 1204–5. [**57, 58**]

ROURKE, A. W. and HEYWOOD, S. M. (1972). Myosin synthesis and specificity of eukaryotic initiation factors. *Biochemistry*, **11**: 2061–2066. [**49**]

SCHIMKE, R. T. and DOYLE, D. (1970). Control of enzyme levels in animal tissues. *Ann. Rev. Biochem.*, **39**: 929–76. [**39**]

——, RHOADS, R. E., PALACIOS, R., and SULLIVAN, D. (1973). Ovalbumin mRNA, complementary DNA, and hormone regulation in chick oviduct. *Karolinska Symposia* in Reproductive Endocrinology, **6**: 357–379 [**81**]

SCHNETTER, W. (1967). Transplantation von Furchungs- und Blastodermkernen in entkernte Eier bei *Leptinotarsa decemlineata* (Coleoptera). *Zool. Anz.*, *Suppl.*, **30**: 494–9. [**16**]

SCHUBIGER, M. and SCHNEIDERMAN, H. A. (1971). Nuclear transplantation in *Drosophila melanogaster*. *Nature*, **230**: 185–6. [**27**]

SCHWARTZ, M. C. (1970). Nucleic acid metabolism in oocytes and embryos of *Urechis caupo*. *Devel. Biol.*, **23**: 241–60. [**78**]

SCONZO, G. and GIUDICE, G. (1971). Synthesis of ribosomal RNA in sea urchin embryos. V. Further evidence for an activation following the hatching blastula stage. *Biochim. Biophys. Acta*, **254**: 447–451. [**78, 80**]

——, PIRRONE, A. M., MUTOLO, V., and GIUDICE, G. (1970). Synthesis of ribosomal RNA during sea urchin development. III. Evidence for an activation of transcription. *Biochim. Biophys. Acta*, **199**: 435–40. [**80**]

SHIOKAWA, K. and YAMANA, K. (1967). Inhibitor of ribosomal RNA synthesis in *Xenopus laevis* embryos. *Devel. Biol.*, **16**: 389–406. [**103**]

SIGNORET, J., BRIGGS, R., and HUMPHREY, R. R. (1962). Nuclear transplantation in the Axolotl. *Devel. Biol.*, **4**: 134–64. [**25**]

SINGER, R. H. and PENMAN, S. (1972). Stability of HeLa cell mRNA in actinomycin. *Nature*, **240**: 100–2. [**68**]

SLATER, D. A., SLATER, I., and GILLESPIE, D. (1972). Post-fertilisation synthesis of polyadenylic acid in sea urchin embryos. *Nature*, **240**: 333–7. [**42**]

SMITH, L. D. (1965). Transplantation of the nuclei of primordial germ cells into enucleated eggs of *Rana pipiens*. *Proc. natn. Acad. Sci.*, *U.S.A.*, **54**: 101–7. [**29**]

—— (1966). The role of a 'Germinal plasm' in the formation of primordial germ cells in *Rana pipiens*. *Devel. Biol.*, **14**: 330–47. [**87, 90**]

—— and ECKER, R. E. (1970). Regulatory processes in the maturation and early cleavage of amphibian eggs. *Curr. Top. in devel. Biol.*, **5**: 1–38. [**8**]

SMITH, M., STAVNEZER, J., HUANG, R.-C., GURDON, J. B., and LANE,

C. D. (1973). Translation of messenger RNA for mouse immunoglobulin light chains in living frog oocytes. *J. mol. Biol.*, **80**: 553–557. [**58, 72**]

SPIRIN, A. S. (1966). On 'masked' forms of messenger RNA in early embryogenesis and in other differentiating systems. *Curr. Top. in devel. Biol.*, **1**: 1–38. [**46**]

STAVNEZER, J. and HUANG, R-C. C. (1971). Synthesis of a mouse immunoglobulin light chain in a rabbit reticulocyte cell-free system. *Nature (Lond.) New Biol.*, **230**: 172–6. [**50**]

STEVENS, R. H. and WILLIAMSON, A. R. (1972). Specific IgG mRNA molecules from myeloma cells in heterogeneous nuclear and cytoplasmic RNA containing poly-A. *Nature*, **239**: 143–6. [**58, 72**]

STEWARD, F. C. (1970). From cultured cells to whole plants: the induction and control of their growth and differentiation. *Proc. R. Soc. B*, **175**: 1–30. [**12, 13**]

SULLIVAN, D., PALACIOS, R., STAVNEZER, J., TAYLOR, J. M., FARAS, A. J., KIELY, M. L., SUMMERS, N. M., BISHOP, J. M., and SCHIMKE, R. T. Synthesis of a deoxyribonucleic sequence complementary to ovalbumin messenger ribonucleic acid and quantification of ovalbumin genes. *J. Biol. Chem.*, **248**, 7530–39. [**9**]

SUSSMAN, M. (1966). Some genetic and biochemical aspects of the regulatory program for slime mold development. *Curr. Top. devel. Biol.*, **1**: 61–83. [**41**]

SUZUKI, Y., GAGE, L. P., and BROWN, D. D. (1972). The genes for silk fibroin in *Bombyx mori*. *J. mol. Biol.*, **70**: 637–49. [**9, 81**]

TATA, J. R. (1971). Protein synthesis during amphibian metamorphosis. *Curr. Top. devel. Biol.*, **6**: 79–110. [**110**]

THOMPSON, L. R. and MCCARTHY, B. J. (1968). Stimulation of nuclear DNA and RNA synthesis by cytoplasmic extracts *in vitro*. *Biochem. Biophys. Res. Commun.*, **30**: 166–72. [**102**]

TIEDEMANN, H. (1967). Biochemical aspects of primary induction and determination. In *The biochemistry of animal development* (ed. R. Weber), **2**: 4–57. [**109**]

TOBLER, H., SMITH, K. D., and URSPRUNG, H. (1972). Molecular aspects of chromatin elimination in *Ascaris lumbricoides*. *Devel. Biol.*, **27**: 190–203. [**12**]

TOMKINS, G. M., GELEHRTER, T. D., GRANNER, D., MARTIN, D., SAMUELS, H. H., and THOMPSON, E. B. (1968). Control of specific gene expression in higher organisms. *Science*, 1474–80. [**51**]

TUOHIMAA, P., SEGAL, S. J., and KOIDE, S. S. (1972). Induction of avidin synthesis by RNA obtained from chick oviduct. *Proc. natn. Acad. Sci., U.S.A.*, **69**: 2814–17. [**54**]

TYLER, A. (1967). Masked messenger RNA and cytoplasmic DNA in relation to protein synthesis and processes of fertilisation and determination in embryonic development. *Symp. Soc. devel. Biol.*, **26**: 170–226. [**46**]

URSPRUNG, H. (1968). (ed.) *The stability of the differentiated state.* Springer Verlag, Berlin. [**14**]

VASIL, V. and HILDEBRANDT, A. C. (1965). Differentiation of tobacco plants from single isolated cells in microcultures. *Science*, **150**: 889–92. [13]

WADA, K., SHIOKAWA, K., and YAMANA, K. (1968). Inhibitor of ribosomal RNA synthesis in *Xenopus laevis* embryos. I. Changes in activity of the inhibitor during development and its distribution in early gastrulae. *Exp. Cell Res.*, **52**: 252–60. [103]

WADDINGTON, C. H. (1956). *Principles of development*. George Allen and Unwin, London. [109]

WALLACE, H. R. (1963). Nucleolar growth and fusion during cellular differentiation. *J. Morphol.*, **112**: 261–78. [20]

—— and BIRNSTIEL, M. L. (1966). Ribosomal cistrons and the nucleolus organiser. *Biochim. Biophys. Acta*, **114**: 296–310. [20]

WANG, S. R., GIACOMONI, D., and DRAY, S. (1973). Physical and chemical characterisation of RNA incorporated by rabbit spleen cells. *Exp. Cell Res.*, **78**: 15–24. [54]

WEBER, R. (1967). Biochemistry of amphibian metamorphosis. In *The biochemistry of animal development*, vol. II, pp. 227–95. Academic Press, New York. [110]

WENSINCK, P. C. and BROWN, D. D. (1971). Denaturation map of the ribosomal DNA of *Xenopus laevis*. *J. mol. Biol.*, **60**: 235–47. [6]

WHITELEY, A. H., McCARTHY, B. J., and WHITELEY, H. R. (1966). Changing populations of messenger RNA during sea urchin development. *Proc. natn. Acad. Sci.*, *U.S.A.*, **55**: 519–25. [81]

WIGLE, D. T. and SMITH, A. E. (1973). Specificity in initiation of protein synthesis in a fractionated mammalian cell-free system. *Nature (Lond.) New Biol.*, **242**: 136–40. [49]

WILSON, E. B. (1904). Experimental studies on germinal localisation. *J. exp. Zool.*, **1**: 197–245. [86]

—— (1925). *The cell in development and heredity*, Macmillan. See especially Chapter XIV. [86]

WILSON, I. B., BOLTON, E., and CUTTLER, R. H. (1972). Preimplantation differentiation in the mouse egg as revealed by microinjection of vital markers. *J. Embryol. exp. Morph.*, **27**: 467–79. [118]

WOODLAND, H. R., and GRAHAM, C. F. (1969). RNA synthesis during early development of the mouse. *Nature*, **221**: 327–32. [78]

—— and GURDON, J. B. (1968). The relative rates of synthesis of DNA, sRNA and rRNA in the endodermal region and other parts of *Xenopus laevis* embryos. *J. Embryol. exp. Morph.*, **19**: 363–85. [103]

—— —— (1969). RNA synthesis in an amphibian nuclear-transplant hybrid. *Devel. Biol.*, **20**: 89–104. [96, 97]

—— and PESTELL, R. Q. W. (1972). Determination of the nucleoside triphosphate contents of eggs and oocytes of *Xenopus laevis*. *Biochem. J.*, **127**: 597–605. [101]

——, FORD, C. C., and GURDON, J. B. (1971). Studies on genetic regulation utilising microinjection of nuclei and DNA into living eggs and oocytes. *Advances in Biosciences*, **8**: 207–18. Pergamon Press, Oxford. [**101**]

WRIGHT, D. A. and MOYER, F. H. (1966). Parental influences on lactate dehydrogenase in the early development of hybrid frogs in the genus *Rana*. *J. exp. Zool.*, **163**: 215–30. [**83**]

WRIGHT, T. R. F. (1970). The genetics of embryogenesis in *Drosophila*. *Adv. Genet.*, **15**: 261–395. [**85**]

YAMADA, T. (1967*a*). Cellular and subcellular events in Wolffian lens regeneration. *Curr. Top. devel. Biol.*, **2**: 249–83. [**15**]

—— (1967*b*). Factors of embryonic induction. In *Comprehensive biochemistry*, **28**: 113–14 (ed. M. Florkin and E. H. Stotz). Elsevier, Amsterdam. [**109**]

ZALOKAR, M. (1971). Transplantation of nuclei in *Drosophila melanogaster*. *Proc. natn. Acad. Sci., U.S.A.*, **68**: 1539–41. [**27**]

ZETSCHE, K., GRIENINGER, G. E., and ANDERS, J. (1970). Regulation of enzyme activity during morphogenesis of nucleate and enucleate cells of *Acetabularia*. In *Biology of Acetabularia* (ed. J. Bonotto and J. Brachet), pp. 87–110. Academic Press, New York. [**40**]

Subject Index